Newer Concepts and Applications for FOOD INDUSTRY

Newer Concepts and Applications for FOOD INDUSTRY

Editor
Y.S. Reddy

2006
Gene-Tech Books
New Delhi - 110 002

ISBN 81-89729-09-8

Published by : GENE-TECH BOOKS
4762-63/23, Ansari Road, Darya Ganj,
NEW DELHI - 110 002
Phone: 41562849
e-mail: genetechbooks@yahoo.co.in

Printed at : Tarun Offset Printers
Delhi

PRINTED IN INDIA

Preface

"We are what we eat" is an old proverb. Our nutritional status, health, physical and mental faculties depend on the food we eat and how we eat it. Access to good quality food has been man's main endeavour from the earliest days of human existence. Safety of food is a basic requirement of food quality. While the world has been changing over the last years both politically and economically in unexpected and remarkable ways, food security remains an unfulfilled dream for currently more than 800 million people.

Beginning with Louis Pasteur's work with wine, modern food science and technology has made enormous contributions to the safety and availability of food. The spectrum of potential benefits from the application of genetic engineering and biotechnology to food crops ranges from diagnostic aids, for example in plant diseases, through to gene mapping, where the genetic characteristics of plants are visibly cartographed, enabling speedier identification of interesting genetic material for every kind of plant usable in agriculture.

Modern biotechnology has enabled scientists to genetically modify plants, animals and micro-organisms rapidly. It can also introduce a greater diversity of genes into organisms – including genes from unrelated species – than traditional methods of breeding and selection.

Although modern biotechnology has demonstrated its utility, there are concerns about the potential risks posed by genetically modified organisms. Today, most countries with biotechnological industries have sophisticated legislation in place intended to ensure the safe transfer, handling, use and disposal of those organisms and their products. Despite these initiatives, the use and development of modern food biotechnology remains a controversial global issue.

This book describes the new concepts and applications for food industry. It analyses the overall contribution of modern biotechnology to food safety and security. It includes the application of modern food biotechnology to microorganisms, plants and animals. An integrated approach was adopted to identify the key issues impacting directly or indirectly on food industry.

Editor

Contents

1

Introduction to Food Industry

The food industry is the complex, global collective of diverse businesses that together supply most of the calories consumed by the world population. Only subsistence farmers, those who eat only what they grow, can be considered outside of the scope of the modern food industry. The industry includes:

— *regulation*: local, regional, national and international rules and regulations for food production and sale, including food quality and food safety, and industry lobbying activities

— *education*: academic, consultancy

— *research and development*: food technology

— *manufacturing*: agrichemicals, seed, farm machinery and supplies, agricultural construction, etc.

— *agriculture*: raising of crops and livestock, seafood

— *food processing*: preparation of fresh products for market, manufacture of prepared food products

— *marketing*: promotion of generic products (e.g. milk board), new products, public opinion, through advertising, packaging, public relations, etc

— *wholesale and distribution*: warehousing, transportation, logistics

— *retail*: supermarket chains and independent food stores, direct-to-consumer, restaurant, food services

In 1865, at the end of the Civil War, farmers made up about 55% of the workforce in the United States. By 1900, 38% of working Americans still toiled on 5.7 million farms-growing enough food to feed the nation's population of 76 million. Today, only about 2.5% of the U.S. workforce is employed on farms. The total number of American farms is down to a little over 2 million, but that dwindling count of farms and farmers meets the domestic needs of a national population approaching 300 million-nearly four times the population of 1900.

Since the early 1900s, the amount of manpower required to grow food has plummeted. The relative cost of an American family's food has likewise dropped impressively. According to the Federal Reserve Bank of Dallas, in 1901 46.4% of a typical American household's income went to food. By 1995, that ratio had dropped to 14.0%.

In 1919, at the end of World War I, a basket of staple food items (one pound each of coffee, bacon, bread, beans, onions, lettuce and ground beef, plus generous amounts of sugar, tomatoes and other items) cost what an average American would earn in 10 hours of work. By 1995, that cost had dropped to less than two hours. The drop has been caused by increases in total personal income, as well as improvements in food technologies. Outside the U.S., other industrialised nations have made outstanding strides in food cost, availability and quality.

Many developing nations have seen vast improvements as well. Meanwhile, throughout much of the world, technology and globalisation have revolutionised the way that we grow food, as well as the way that we transport, process, package, purchase and cook it. Waste and spoilage are down to nominal levels thanks to innovations like interstate highways and refrigerated trucks.

Furthermore, it's an everyday occurrence for consumers in the U.S. to pick up strawberries from New Zealand or mangos from Mexico in the fresh produce section of the local supermarket. Globalisation has led to the rise of massive multinational food processing companies like Nestlé and Kraft, which often sell their foods under local names in local languages, after producing them in regional factories worldwide.

The types of technologies affecting the food industry have evolved over time. From mechanised tractors and implements to diesel trucks to flash freezing, food technology has moved on to become high-tech. Today, computerisation has made marked changes in the food industry: Electronic data interchange ensures that inventories and shipments are well managed so your local grocer doesn't run out of the products that are selling quickly. Point-of-sale systems at the cash register capture minute-by-minute sales data.

Biotechnology is making sweeping changes at the ground level-in seed stocks and agricultural animal health. In fact, gradual genetic improvement of grain seeds like rice and wheat, combined with better fertilisers and other technologies, has created a "green revolution," enabling nations like China and India to go from agonizingly underfed populations to a large degree of food self-sufficiency and, in some cases, net exports of bounteous crops. Now, genetically modified seeds are gaining ground with the promise of crops that not only resist insects and have extremely high yields per acre, but also produce high levels of desirable nutrients and vitamins.

Growing health concerns are significantly impacting all sectors of the food industry as obesity levels continue to rise to alarming proportions in the U.S. and elsewhere. The Federal Government, the Food and Drug Administration (FDA) and a host of consumer groups are squaring off with food producers over nutrition and the responsibilities and ethical issues inherent in the production and marketing of food.

Milk Processing

Milk fresh from the cow is virtually a sterile product. All post-milking handling must maintain the milk's nutritional value and prevent deterioration caused by numerous physical and biological factors. In addition, equipment on the farm must be maintained to government and industry standards. Most cows are milked twice a day, although some farms milk three or four times per day.

The milk is immediately cooled from body temperature to below 40°F (5°C), then stored at the farm under refrigeration until picked up by insulated tanker trucks at least every other day. The milk tanker driver records the amount of milk and notes the temperature and the presence of any off-odors. If the milk is too warm or has an off-odor, it will not be picked up, and the farmer will have to feed it to his animals or dump it. When the milk is pumped into the tanker, a sample is collected for later lab analysis.

When the milk arrives at the milk plant, it is checked to make sure it meets the standards for temperature, total acidity, flavor, odor, tanker cleanliness, and the absence of antibiotics. The butterfat and solids-not-fat content of this raw milk is also analysed. The amounts of butterfat (BF) and solids-not-fat (SNF) in the milk will vary according to time of year, breed of cow, and feed supply.

Butterfat content, solids-not-fat content, and volume are used to determine the amount of money paid the farmer.

Once the load passes these receiving tests, it is then pumped into large refrigerated storage silos (nearly half-million pounds capacity) at the processing plant. All raw milk must be processed within 72 hours of receipt at the plant. Milk is such a nutritious food that numerous naturally occurring bacteria are always present. The milk is pasteurised, which is a process of heating the raw milk to kill all "pathogenic" bacteria that may be present. A pathogen is a bacteria that could, if allowed to grow and multiply, make humans sick. It should be noted that pasteurisation is not sterilisation (sterilisation eliminates all viable life forms, while pasteurisation does not).

After pasteurisation, some harmless bacteria may survive the heating process. It is these bacteria that will cause milk to "go sour." Keeping milk refrigerated is the best way to slow the growth of these bacteria. Some bacteria do not cause spoilage, but are actually added to milk or cream after pasteurisation to make "cultured" products such as cheese, cottage cheese, yogurt, buttermilk, acidophilus milk and sour cream.

There are different ways to pasteurise milk. The "batch" method heats the milk to at least 145° and holds it at that temperature for at least 30 minutes. Since this method may cause a "cooked" flavor, it is not used by some milk plants for fluid milk products. High Temperature/Short Time (HTST) pasteurisation heats the milk to at least 161° for at least 15 seconds. The milk is immediately cooled to below 40° and packaged into plastic jugs or plastic-coated cartons. Most milk plants have at least one HTST processor. This piece of equipment is considered the "heart" of the plant.

Butterfat content accounts for several different types of products. Whole milk, 2%, 1%, Nonfat, and Half & Half are some examples. A machine called a separator separates the cream and skim portions of the milk. A separator is really a large centrifuge that spins about 2,000 rotations per minute. The different types of milk products are then "standardised" by blending the components (skim milk, raw milk, cream) in the correct proportions to yield the desired end-products. Water is never added to lower the butterfat content of fluid milk. Excess cream is used to make ice cream and butter.

Milk is homogenised to prevent the cream portion from rising to the top of the package. The expression "cream rises to the top," is accurate because cream is lighter in weight than milk. The cream portion of un-homogenised milk would form a cream layer at the top of the carton. A "homogeniser" forces the milk under high pressure through a valve that breaks up the butterfat globules to such small sizes they will not "coalesce" (stick together). Homogenisation does not affect the nutrition or quality of the product; it is done entirely for aesthetic purposes.

Vitamin quantities may be reduced by the heating process and removal of the butterfat. Therefore, to replace the natural nutrition of nature's perfect food, liquid vitamins are added to fortify most fluid milk products. Many states have milk standards that require the addition of milk solids. These solids represent the natural mineral (i.e. calcium, iron), protein (casein), and sugar (lactose) portion of nonfat dry milk.

Quality Control personnel conduct numerous tests on the raw and pasteurised products to insure optimum quality and nutrition. A sample is analysed for the presence of microbiological organisms with a standard plate count (SPC) and ropey milk test. The equipment used to analyse butterfat and solids-not-fat is calibrated on a regular basis to insure a consistent, quality product that meets or exceeds government requirements.

All milk products have a sell-by date printed on the package. This is the last day the item should be offered for sale. However, most companies guaranty the quality and freshness of the product for at least 7 days past the date printed on the package. Samples of each product packaged each day are saved to confirm that they maintain their freshness 7 days after the sell-by date.

Once the milk has been separated, standardised, homogenised and pasteurised, it is held below 40°F in insulated storage tanks, then packaged into gallon, half-gallon, quart, pint, and half-pint containers. The packaging machines are maintained under strict sanitation specifications to prevent bacteria from being introduced into the pasteurised product. All equipment that comes into contact with product (raw or pasteurised) is washed daily. Sophisticated automatic Clean-in-Place (CIP) systems guarantee consistent sanitation with a minimum of manual handling, reducing the risk of contamination.

Once packaged, the products are quickly conveyed to a cold storage warehouse. They are stored there for a short time and shipped to the supermarket on refrigerated trailers. Once at the store, the milk is immediately placed into a cold storage room or refrigerated display case.

Food Safety

Food safety is a very broad topic. Pesticides, herbicides, chemical additives, and spoilage are all of concern, but food scientists, food processors, and consumers focus most on microbiological quality. Microorganisms pose a challenge to the food industry and most food processes are designed with microbial quality in mind. Microorganisms are often too small to be seen with the unaided eye and have the ability to reproduce rapidly. Many of them produce toxins and can cause infections.

For all of these reasons, the microbiological quality of the food we eat is scrutinised closely.

Centuries ago, Genghis Khan was able to rule vast stretches of land through the mobility of his army. With very little food, he was able to engage in swift attacks over long periods of time. As the story goes, each horseman carried two leather bags. The larger one held dry milk produced by drying fluid milk in the sun during periods of rest. The smaller bag was used to rehydrate some of the milk powder with water, which was consumed during an offensive. The lightly equipped army of Khan thus could cover long distances in weeks, and eventually controlled most of the Asian continent. Yet, one has to wonder how many people suffered food-borne illness in those days.

Today, food-borne illness is of serious concern. Its frequency is not known because a great majority of the cases go unreported. Reporting food-borne diseases to public health authorities is not required in the United States. Estimates claim as many as 200 million cases in the U.S. per year. Only a small percentage of these are hospitalised. Most are passed off as traveler's diarrhea, 24 hour flu, or upset stomach. Salmonellosis, one of the more serious food-borne diseases, is said to be reported only about 1% of the time. About 42,000 cases of salmonellosis are reported in the U.S. annually, with about 150 deaths. So, there are potentially 4.2 million cases of Salmonella food poisoning annually despite the fact that the U.S. food supply is considered very safe and processed under the best conditions available.

Testing the foods we consume for the presence of pathogenic microorganisms is very important. Although 100% of the food cannot be tested, it can be deemed "safe" through proper audit of the food supply. In many instances, the pathogenic microorganisms are present in very small numbers, but for many of these pathogens, small numbers are all that are necessary to transmit disease or illness. For that reason, the presence of other microorganisms is monitored. These microorganisms provide an index of the sanitary quality of the product and may serve as an indicator of potential for the presence of pathogenic species. Escherichia coli (E. coli) is commonly employed as an indicator microorganism. Because E. coli is a coliform bacteria common to the intestinal tract of humans and animals, its relationship to intestinal food-borne pathogens is high.

Total counts of microorganisms are also an indication of the sanitary quality of a food. Referred to as the Standard Plate Count (SPC), this total count of viable microbes reflects the handling history, state of decomposition or degree of freshness of the food. Total counts may be taken to indicate the type of sanitary control exercised in the production, transport, and storage of the food. Most foods have standards or limits for total counts. This is especially true for milk.

It must be remembered that a low SPC does not always represent a safe product. It is possible to have low-count foods in which toxin-producing organisms have grown. These organisms produce toxins that remain stable under conditions that may not favor the survival of the microbial cell. In adopting microbiological standards to milk, the first concern is product safety, followed by shelf-life. The following bacterial counts are standards for milk as recommended by the U.S. Public Health Service:

— Grade A raw milk for pasteurisation Not to exceed 100,000 bacteria per milliliter (ml) prior to commingling with other produced milk; and not exceeding 300,000 per ml as commingled milk prior to pasteurization.

— Grade A pasteurised milk Not over 20,000 bacteria per ml, and not over 10 coliforms per ml.

The objective of pasteurisation is to reduce the total microbial load, or SPC. In addition, pasteurisation must destroy all pathogens that may be carried in the milk from the cow, particularly undulant fever, tuberculosis, Q-fever, and other diseases transmittable to humans. This is accomplished by setting the time and temperature of the heat treatment so that certain heat-resistant pathogens, specifically Mycobacterium tuberculosis and Coxiella burnetii (causative agents of Q-fever and tuberculosis, respectively) would be destroyed if present. Milk pasteurisation temperatures are sufficient to destroy all yeasts, mold, and many of the spoilage bacteria.

Quality Assurance

Of all functions in the food industry, Quality Assurance (QA) requires many diverse technical and analytical skills. QA personnel continually monitor incoming raw milk and finished milk products to insure compliance with compositional standards, microbiological standards, and various government regulations. A QA manager can halt production, refuse acceptance of raw material, or stop the shipment if specifications for a product or process are not met. This department does not usually have control over the product unless something has gone wrong.

The major functions of the QA Department are:

— Compliance with specifications Legal requirements, industry standards, internal company standards, shelf-life tests, customers' specifications.

— Test procedures Testing of raw materials, finished products, and in-process tests.

— Sampling schedules Utilise a suitable sampling schedule to maximise the probability of detection while minimising workload.

— Records and reporting Maintain all QA records so that customer complaints and legal problems can be dealt with.

— Trouble shooting Solve various problems caused by poor quality raw materials, erratic supplies and malfunctioning process equipment; and investigate reasons for poor quality product to avoid repetition.

— Special problems Customer complaints, production problems, personnel training, short courses, etc.

A typical QA Department may have a chemistry lab, a raw materials inspection lab, a sensory lab and a microbiology lab. All of these disciplines work together to assure that 'the food we consume is of the highest quality. After all, it is quality which will bring a customer back again and again.

Career Opportunities

There are a number of career opportunities in Quality Assurance. Typically, these positions are held by graduates of university food science programs. In a typical hierarchy, technicians report to a laboratory supervisor who in turn reports to the QA manager. In some food companies, the technicians do not necessarily have a college degree. With the widespread use of more specific, high-tech analytical techniques, however, it is becoming more important for a QA technician to have a minimum 2-year college degree emphasising the sciences.

Examples of QA personnel careers are:

— Microbiology lab technician Responsible for the day to day activities in the microbiology lab.

Knowledgeable in sterile technique, mathematics and accurate reporting of data.

- — Chemistry lab technician Responsible for the day to day chemical analysis of food in the chemistry lab. Knowledgeable in basic chemistry, physics, mathematics and accurate reporting of data.
- — Inspection lab technician Knowledge of the sanitary handling of food samples, mathematics and record-keeping.
- — Customer service Responsible for handling customer complaints and tracking down test results and/or product. Knowledge of specifications and broad legal responsibilities. Helps to be people-oriented.
- — Specifications Responsible for accurate, organised record-keeping so that, if needed, product can be traced and verified.
- — Laboratory supervisor In charge of the lab personnel and their daily activities. Responsible for scheduling both personnel and testing.

Food Processing

Making Peanut Butter

All of the peanuts used for peanut butter are grown in the USA, either in the Southeast or the Southwest. The two best-tasting varieties for peanut butter are Spanish and Runners (so Safeway used only these two varieties). From the harvest the nuts go to shelling operations. These plants, located near the growing fields, remove the shells, clean the nuts, and pack them into huge bags for shipment to the peanut butter plant. Each bag holds more than 2,000 pounds of peanuts!

At the plant, the bags are unloaded into bucket conveyors that move the nuts from each processing step to the next one. The first step is to insure that all impurities, such as stems and sticks from the peanut plants, are removed from the product stream. This is done by gravity separators, which sort out objects that are heavier or lighter than peanuts.

The peanuts are now ready for roasting. This is done by a continuous roaster. The nuts are slowly carried through the roaster on a belt while hot air is circulated. It is extremely important that the nuts be roasted evenly and properly so that flavor and color are just right. The roaster operator adjusts the roast as required by changing the air temperature, belt speed or peanut layer thickness on the belt. The peanuts are cooled and conveyed to the blanching machines, which remove the skins. This prevents the peanut butter from having dark specks from the skins.

The last step before the peanuts can be ground into peanut butter is the final inspection for quality. The nuts are conveyed through an electronic color sorter which removes nuts that were under or over roasted. The peanuts also pass a trained inspector who looks them over and picks out any that do not look right.

The peanuts are now ready to be conveyed to the grinders. The grinders are like giant milkshake machines. Although peanut butter consists of mostly peanuts (at least 90%), small amounts of other ingredients are added while the nuts are being ground. In the case of 'old fashioned' peanut butter, we add a little bit of salt for flavor and a special vegetable oil called a stabiliser. This keeps the peanut oil from separating out to the top of the jar.

At this point the peanut butter is pumped through a metal detector to insure that no metal got into it during the grinding. Then it is pumped into a deaerator, which removes trapped air. Finally, the peanut butter, quite hot from all that grinding, passes through a heat exchanger to cool it down so it can be packed into containers on the filling line. The filling machine is carefully timed to put the correct amount of peanut butter in each jar. The jar then is conveyed to the capping machine.

Capped jars are sent through an induction sealer, which seals the inner liner to the top of the jar. Then, another machine applies the label to the jar. The jars are now ready for packing into the shipping case. This is done by hand so that the packer can inspect each jar's label and general appearance.

All that's left is to glue the cases closed and put the peanut butter in the warehouse, then wait a few days before shipping it to the stores. This allows for microbiological tests to make sure no molds or bacteria have found their way into the peanut butter. After the tests come back 'clean' we can release it for our customers to enjoy.

Making Bread

Making bread involves many ingredients and advance preparation steps. Flour is received by bulk rail or truck and stored in 100,000 pound bins. All other raw materials are received by truck. Two hours before production begins, a liquid sponge or broth is prepared and allowed to ferment to ensure that the finished loaf will rise properly. The broth is a blend of flour, water, sugar, salt, yeast and yeast foods.

To combine the ingredients necessary for bread making, a scaler measures out the smaller increments of the mix, some as little as one ounce. A dough mixer operated by a control panel takes the ingredients from the scaler and adds the larger increments to the mix to create the proper dough consistency.

This mixture can weigh anywhere from 400 to 2,000 pounds. The dough is then 'kicked' out of the mixer into a trough and allowed to 'relax' and ferment. This is called floortime. Then it goes to a hopper and is divided into loaf-sized pieces, then to the rounder for shaping.

Once again the dough is set aside in an overhead proofer to relax and continue fermenting for approximately 10 minutes. The dough is then sent to the head rollers for flattening and removal of excess air. This is a key step in bread making. Removing excess fermenting gas helps ensure good inner structure and grain in the finished loaf.

The next stop is the moulder, where the bread is shaped for the final baking process. The moulder is also helpful in removing air from the dough. Once molded, the bread is dropped into a large pan divided into five separate loaf pans. These pans travel along a conveyor to another proof box. Here they will stay for 55 minutes. The temperature in the proof box is monitored closely to maintain 90% humidity level and 105° temperature level at all times.

Now the bread is sent to the ovens for baking. The oven temperatures and baking times will vary as to size and density of the loaf. The loaves bake for 22 minutes at approximately 400°. The baked bread is conveyed to a depanner. This is just what it sounds like; suction cups and vacuum pressure remove the baked loaf from the pan. The pan is sent back to storage to be used again, and the loaf is sent to cool.

The bread cools for about an hour and is then sent to be sliced. Once sliced, the bread is wrapped by an automatic bagging machine. Now that the loaf is in the bag, it is sent to be tied and fastened. The finished product is conveyed to where it is sorted and stacked for store distribution. Total production time for a loaf of bread is about three hours. The total lapsed time from the beginning of production to when the bread is on the shelf in the store is 24 hours.

Making Strawberry Jam

The fruit preserves you buy in supermarkets are the product of many dedicated people whose efforts are coordinated into a complex team effort. The process or recipe for making preserves is fairly simple and has been done in home kitchens for many years. Through more than 30 years of experience in the application of food science and processing engineering, Processors/marketers, such as Safeway, have been able to develop large-scale processing methods to improve quality and consistency found in the home kitchen.

The starting point is the farmer's field. The coastal valleys of California, Oregon and Washington produce the finest fruit in the world. Safeway buys the best of this crop from the premier growers of the region. Fruit is hand picked in the fields and quickly transported to packing houses where it is cleaned and quick frozen to maintain color, flavor, texture and to prevent spoilage. Properly frozen fruit, when held under the right conditions, will maintain its quality for several years. However, the freeze/thaw cycle does cause some damage to the fruit regardless of how carefully the process is done.

The best preserves are made with fresh fruit (never frozen). The fruit is chilled and shipped directly from the packing house to the preserve plant where it is processed within 24 hours. In addition to buying only from premier growers and packing houses, field inspectors regularly visit the packing houses to ensure that the fruit meets the quality standards.

The frozen fruit is held in cold storage warehouses. These warehouses, which are gigantic freezers, keep the fruit at the proper temperature to insure that it maintains its optimum quality. Just prior to cooking and packaging the fruit is shipped by truck to the preserves plant, inspected by trained food chemists and thawed under careful observation to the proper point.

As stated earlier, the starting point is the field, but the real beginning is when the marketing manager makes the decision as to the number and kinds of preserves to pack. This decision is made from 18 to 20 months in advance and is the result of many hours of research and consultation. This advance forecasting of future demand must be done so that the purchasing agent can make buying commitments during or prior to the harvest.

Within one to two weeks of the actual pack date, the plant's production control manager will make a detailed schedule of which products will be produced on any given day. At that time he and his staff will issue purchase and release orders for the elements needed to produce the finished product such as: jars, caps, labels, glue, etc. It is the production control manager's responsibility to have all the items needed in the plant at the right time.

Once the production schedule has been made, the plant operations people meet and review the schedule in detail. This group includes the departmental foreman, the maintenance manager, the quality control manager, the production control manager, the production manager and supervisors. It is at this meeting that the final details are decided on and all the activities of the various departments are coordinated.

From three to five days prior to production, the fruit is received at the plant, inspected by the quality control food chemist and set out to thaw. The fruit is thawed to the point that it can be managed but is still at the freezing temperature of 32° F. Once the thawing process has begun, the packaging process is committed. The fruit cannot be re-frozen without harm.

The commercial production of preserves uses two to three kinds of sugar, depending on the variety. Different kinds of liquid sugar and dry sugar may be used. The sugars are received in bulk by truck and stored in tanks and silos. In addition to the fruit and sugars, a small amount of pectin is used to provide the set to the liquid portion of the preserve. The original preserve recipes used gelatin, an animal product. However, gelatin has been replaced by pectin, a natural fruit product, from which the term jelly comes. A small amount of citric acid, another natural fruit product, is used to adjust the acidity for flavor consistency.

Nutritional Food

Nutrition is the process by which the foods we eat provide the nutrients we need to grow and stay healthy. Nutrients are naturally occurring chemical substances found in food. There are six categories of nutrients: proteins, lipids, carbohydrates, vitamins, minerals and water.

Proteins contain amino acids, sometimes referred to as the building blocks of protein. Dietary protein is supplied from plant and animal sources. Proteins are needed to build and repair body tissue and for the metabolic functions of our bodies.

Lipids include fats and oils from plants and animals. Cholesterol is a fat found only in animal products. Lipids are of special interest because they are linked to the development of heart disease, the leading cause of death among Americans.

The carbohydrates in our diet come from plant foods. Simple carbohydrates include the different forms of sugar, while complex carbohydrates include starches and dietary fiber. Vitamins are chemical compounds in our food that are needed in very small amounts (in milligrams and micrograms) to regulate the chemical reactions in our bodies.

Minerals, also needed only in small amounts, have many different functions. Some minerals assist in the bodyís chemical reactions and others help form body structures. Fifty to sixty percent of our body weight consists of water. It is the substance in which the metabolic reactions occur. We need about two quarts (2 liters) of water every day.

Protein, fats and carbohydrates in food provide the energy, or kilocalories (kcals), our bodies need to function. Each gram of protein and carbohydrate has 4 kilocalories; each gram of fat has 9. You might have noticed that we use the metric system - grams, milligrams and micrograms - to measure the amounts of nutrients in foods.

Healthy Diet

"Food guides" are designed to help us learn the types and amounts of foods we need to eat every day for adequate nutrition. These guidelines divide foods into different groups based on the amounts of nutrients the foods contain. Other guidelines, based on the recommendations in the "Dietary Guidelines for Americans," have been developed to help people choose a healthy diet.

In 1992 the U.S. Department of Agriculture released the "Food Guide Pyramid" to replace the "Basic Four Food Groups." The leaflet entitled, *The Food Guide Pyramid: Beyond the Basic,* also is available in the reference library. There is a national nutrition education campaign, co-sponsored by the National Cancer Institute (NCI) and the Produce for Better Health Foundation (PBH), to encourage Americans to eat at least five servings of fruits and vegetables daily. The "5 a Day" health promotion was developed in 1988 by the California Department of Health Services.

Food Labeling

Food labels for most of the food products sold in the United States must have the product name, the manufacturerís name and address, the amount of the product in the package and the product ingredients. The ingredients are listed in descending order, based on their weight. Under the current laws, fresh fruits, vegetables and meat are exempt from these labeling requirements.

In 1973 the Food and Drug Administration (FDA) established "nutrition labeling" or guidelines for labeling the nutrient and caloric content of food products. Nutrition labeling is mandatory only for those foods that have nutrients added or make a nutritional claim. Manufacturers are encouraged, but not required, to provide nutrition labeling of other food products.

The current nutrition labeling regulations from the FDA require a label and have the percentage of the U.S. Recommended Daily Allowances (U.S. RDA). These standards are based on the 1968 edition of the Recommended Dietary Allowances (RDA), but the RDA and the U.S. RDA are not the same! For each nutrient, the U.S. RDA are the highest RDA for any of the RDA age and sex groups. The U.S. RDA usually apply to people four years of age and older.

FDA nutrition labels must have the serving size; servings per container; calories per serving; grams of protein, carbohydrate and fat per serving; and the percent of the U.S. RDA for protein, five vitamins and two minerals. There is less nutrition information on labels regulated by USDA. USDA labels list only the serving size; servings per container; calories per serving; and grams of protein, carbohydrate and fat per serving.

Food labeling provides basic information about the ingredients in, and the nutritional value of, food products so that consumers can make informed choices in the market place. In the 1990 Food Marketing Institute (FMI) survey, over 70% of food shoppers identified taste, nutrition and product safety as being very important factors in making food purchases. In the same survey, 36% of shoppers reported they always read the ingredient and nutrition labels, and another 45% said they sometimes read nutrition labels.

The growing importance of the role of nutrition in promoting health and preventing disease, and consumer demand for clearer and easier to understand information, has led to the passage of the Nutrition Labeling and Education Act (NLEA) of 1990. The federal regulations, detailing the format and content of food labels, are now in effect.

Packaging

Food packaging development started with humankindís earliest beginnings. Early forms of packaging ranged from gourds to sea shells to animal skin. Later came pottery, cloth and wooden containers. These packages were created to facilitate transportation and trade.

Utilizing modern technology, todayís society has created an overwhelming number of new packages containing a multitude of food products. A modern food package has many functions, its main purpose being to physically protect the product during transport. The package also acts as a barrier against potential spoilage agents, which vary with the food product. For example, milk is sensitive to light; therefore, a package that provides a light barrier is necessary.

The milk carton is ideal for that. Other foods like potato chips are sensitive to air because the oxygen in the air causes rancidity. The bags containing potato chips are made of materials with oxygen-barrier properties.

Practically all foods should be protected from filth, microorganisms, moisture and objectionable odors. We rely on the package to offer that protection.

Aside from protecting the food, the package serves as a vehicle through which the manufacturer can communicate with the consumer. Nutritional information ingredients and often recipes are found on a food label. The package is also utilised as a marketing tool designed to attract your attention at the store. This makes printability an important property of a package.

The food industry utilises four basic packaging materials: metal, plant matter (paper and wood), glass and plastic. A number of basic packaging materials are often combined to give a suitable package. The fruit drink box is an example where plastic, paper and metal are combined in a laminate to give an ideal package. This concept can be easily seen in your peanut butter jar. The main package containing the food (primary package) is made of glass (or plastic), the lid is made of metal lined with plastic, and the label is made of paper.

Each basic packaging material has advantages and disadvantages. Metal is strong and a good overall barrier, but heavy and prone to corrosion. Paper is economical and has good printing properties; however, it is not strong and it absorbs water. Glass is transparent, which allows the consumer to see the product, but breakable. Plastics are versatile but often expensive. Therefore, combining the basic materials works well in most cases. So, for a product like milk, which is an essential food for children and young adults and therefore cannot be very expensive, paper makes a good economical material. It also provides a good printing surface

However, since paper absorbs water, it will gain moisture from the milk, get weaker and fail, thereby exposing the milk to spoilage factors. It may even break and waste the product. When a thin layer of a plastic called polyethylene is utilised to line the inside of the milk carton, it serves as a barrier to moisture and makes an economical, functional package.

After making a food product and placing it in the appropriate package, a number of these individual packages must be placed in a large container to facilitate shipment. These larger containers are called secondary packages. The paperboard box is a very common secondary package. Plastics also can serve as secondary packages. The milk case in which a number of milk cartons are delivered to the supermarket is a good example.

Clearly, recycling is a sound approach. However, the problem often lies in feasibility of collection, separation and purification of the consumerís disposed food packages. This mode of recycling is called post-consumer recycling. While it offers a logistic challenge, recycling is gaining in popularity, and the packaging industry is cooperating in that effort. Aluminum cans are the most recycled container at this time. Plastic recycling is increasing, yet most plastic is recycled during manufacturing of the containers; not as post-consumer recycling. For example, trimmings from plastic bottles are reground and reprocessed into new ones.

The plastics industry is helping to facilitate consumer recycling by identifying the type of plastic from which the container is made. A number from 1 to 7 is placed within the recycling logo on the containerís bottom. For example, refers to PET (Polyethylene Terephthalate), the plastic used for the large liter soft drink bottles. Plastics have the advantage of being light. This helps to conserve fuel during transport and also reduces the amount of package waste.

There are many interesting packaging concepts being explored by the industry to keep up with the changing life style of the consumer and new technologies. Many professionals are involved in designing and manufacturing the modern package. Today's package is designed with the consumer's safety and convenience in mind. Examples are microwaveable popcorn packages, squeezable ketchup bottles and the tamper-proof milk bottle cap.

Packaging

To secure a job with the food packaging industry, a college or university degree in food packaging or food science with a packaging minor would be ideal. In today's competitive environment of world-class manufacturing, food processing companies need to manage all resources effectively in order to design and produce the right food products, at the right time, in the right quantities, right the first time. This is what is referred to as integrated resource management.

Computer-based Systems in Manufacturing

Computers are playing an increasingly important role in our lives. Keeping pace with the application of the latest technology is a challenge for all of us. In the food industry, as in manufacturing in general, computers are being put to use designing food products, processing foods and managing the operations and resources of a food processing plant. A manufacturer needs to keep up with the latest technology to remain competitive.

Computer-based systems in manufacturing are known as CAD, CAM,MRPII, and CIM. Although their use is more limited in the food industry than in other manufacturing environments, these concepts are gaining importance rapidly in their application to food processing.

APICS, the American Production and Inventory Control Society, defines the acronyms representing the various computer-based systems as follows:

CAD (Computer-Aided Design) - The use of computers in interactive engineering drawing and storage of designs. Programs complete the layout, geometric transformations, projections, rotations, magnifications and interval (cross-section) views of a part and its relationship with other parts.

CAM (Computer- Aided Manufacturing) - Use of computers to program, direct and control production equipment in the fabrication of manufactured items.

CIM (Computer-Integrated Manufacturing) - The integration of the total manufacturing organisation through the use of computer systems and managerial philosophies that improve the organisation's effectiveness; the application of a computer to bridge various computerised systems and connect them into a coherent, integrated whole. For example, budgets, CAD/CAM, process controls, groups technology systems, MRPII, financial reporting systems, etc., are linked and interfaced.

MRPII (Manufacturing Resource Planning) - A method for the effective planning of all resources of a manufacturing company. MRP II is a computer-based, business information management tool that helps manufacturers produce the "correct" products at the "correct" time. MRP II, used effectively, integrates all company functions in support of overall corporate strategy, utilising accurate, consistent and timely information.

MRP II is an American technology that has evolved gradually over the last 30 years. Benefits to MRP II users include reduced delivery lead times, reduced product shortages, reduced investment in inventories, immediate performance measurements and reporting, faster new product introductions, better cost controls, increased productivity and improved quality of work life. Companies that have been highly successful at implementing MRP II, or the Class A users, all have two things in common: 1)

Detailed education programs for their employees and 2) Employees who are committed to supporting and maintaining the data integrity of the system.

The purpose of this module is to give you a high-level overview of the multiple, interconnecting disciplines that must be joined in a single, people-oriented system of checks and balances to accomplish the MRP II style of management. You will see that MRP II, used effectively, is a top-down management control system using a computer- based model to guide decision making. This model relies on feedback from the bottom up to correct its assumptions and to reflect reality. This process is called closed-loop management.

Classic or traditional manufacturing organisations in the United States have reflected the specialist-oriented educational degree programs offered through American universities and colleges. Students graduate with degrees in engineering, accounting or marketing and go to work as specialists, often for their entire professional career.

A stereotypical career path for a person with a degree in Engineering after joining a traditional American manufacturing firm could include several years as a design Engineer, a promotion to Engineering Supervisor and retirement after becoming the Vice President of Engineering. Specialists in traditional American companies do not communicate effectively with other departments, develop separate systems and goals, and incur an "us against them" attitude within the organisation.

A traditional company with specialist-oriented roles does not foster the team effort needed to become a "world-class" enterprise. By contrast, team effort and support is essential to maintaining the sharedMRP II data base integrity.

Levels of Manufacturing Management

In the MRP II concept of formal management each company should be managed at three levels:

1. Finance and Administration Focus
2. Manufacturing Focus
3. Sales and Marketing Focus

It is essential that the management practices at each of these three levels support one another and, together, implement the business plan. Each management level must carry out its management mission by continuously cycling through the three steps of "plan, execute and measure". Closing the loop consists of using the measurement information from the previous cycle to update the planning assumptions for the next cycle. This is a continuous process for improvement.

Each level of manufacturing management is supported by specific system elements, procedures, policies and computer programs. The computer programs that provide automated support of the three steps by level are:

— *Financial*: Software needs include budgets and allocations (plan); accounts payable and accounts receivable (execute), and general ledger and product costing (measure).

— *Sales and Marketing*: Software requirements for the demand level include forecasting (plan); order entry and billing (execute); and sales analysis (measure).

— *Manufacturing*: Software requirements include master production scheduling, distribution requirements planning, material requirements planning and capacity requirements planning (plan); purchasing and receiving, shop floor control and inventory control (execute); and performance measurement (measure).

Hierarchy of Functions

Most industry experts when discussing MRP II are referring to 10 or 12 discrete business functions from the highest level of planning (business planning) to the lowest level of execution (purchasing & receiving).

Each management planning level receives its vision and constraints from the prior level. Decisions are "pushed down" (top-down) through the hierarchy and performance measurement information is pushed back up (bottom-up) to compare the actual performance with the planned performance. When there is a discrepancy between actual performance and planned performance, replanning is done at the appropriate level. This on-going process of top-down, bottom-up replanning can best be visualised as a circular joining of the three words: Plan, Execute, and Measure.

Issues in Inventory Management

Stores must "inventory" products ahead of sales and "forecast" (or estimate) which products customers will purchase (when and how much) to establish appropriate inventory levels for the various products they carry. What the stores are doing is called inventory management and sales forecasting. Not a simple task, considering that a grocery store carries as many as 30,000 different items, many of which are perishable foods with a very short shelf life. Due to the complexity of managing inventory and forecasting sales, companies are increasingly using computer systems to control their inventory.

Optimal inventory

From the perspective of a retail store, the inventory is the stock of tangible goods to be sold at that store. A store is interested in carrying an "optimal" inventory. Various factors come into place when considering what is an ideal inventory level, such as having the necessary products available in the right quantity at the right time, while minimising the cost of ordering and carrying the goods. Before calculating a store's optimal inventory, we need to give a simple definition of some basic inventory management terms, such as:

Optimal inventory can be computed using the following formula:

$$OI = OP + 1/2\ RQ$$

Where OI = optimal inventory, OP = order point and RQ = replenishment quantity

On the average, all items in the store should be halfway through their replenishment. The goal of any store is to minimise the optimal inventory. Stores can do that by better forecasting their demand (this will lower inventory) and by reducing lead time (this will lower replenishment quantity).

Paths to Ordering

When stores place an order, the finished goods come either directly from the manufacturing plant, or from the store's distribution center or warehouse (which ordered it from a manufacturer). Increasingly, orders are being placed electronically by a computer linking the ordering and supplying parties. The same "order" means two different things depending on who ships or receives the goods. The entity placing the order (in this case, the store) issues a purchase order, the party supplying the merchandise (in this case, the manufacturing plant or the distribution center) considers it a customer order.

When a plant receives an order from a store, it needs to check whether it has enough inventory to ship, or if it needs to produce or manufacture the products. Real perishable goods with short shelf lives, such as bread and milk, are "manufactured to order." This means that the products are processed after the order is placed. Other products are "manufactured to inventory" or made to stock in a warehouse. For example, ice cream is made at Safeway's ice cream plant then transported to the distribution center and stored in freezers. The ice cream subsequently will be shipped to the Safeway stores when they place a replenishment order.

Transportation

When goods are made to inventory, they are usually stocked in a warehouse or distribution center. Whether they are ordered from the plant or the distribution center, the goods need to be transported to the store. A variety of carriers are used to transport the grocery products to the store: trucks, airplanes, trains and boats. Supermarkets may have their own fleet of trucks to transport goods, and supplements its shipping with common air, land and water carriers. They may also have a state-of-the-art distribution center.

Careers

Careers related to inventory management and distribution include people working in the areas of store inventory, order management, distribution, warehousing and transportation.

Marketing

Marketing can be defined as "the activities of a business to acquire knowledge of the market place and improve its performance." The primary measure of performance is satisfying customer needs by meeting, exceeding and anticipating their requirements. If a firm can manage customer satisfaction appropriately (within certain resource and time considerations), it will achieve "profit" as the ultimate result. After all, maximising profits is dear to the shareholders of a company. The business processes and tasks that make up marketing include marketing information, product development, pricing, distribution and promotion.

Marketing Information

A company needs to have timely, accurate, well-organised data so it can make decisions based on information that has a high degree of integrity. The types of information a company is interested in include knowledge about customers and their buying patterns, knowledge about competitors, and knowledge about the market factors affecting the business. Companies obtain that information by using secondary sources (materials that are already published) or by gathering their own primary data (mainly through marketing research).

Safeway is rich in primary data on the movement of competitive food and consumer products in most U.S. markets through the availability of scanner data. Data is obtained on each product that is purchased by a consumer at a Safeway store when the food clerk "scans" the UPC (an acronym for Universal Product Code) at the cash register. Safeway is in a position to use that information strategically to promote its performance in the market place.

Product Development

Many activities at supermarkets relate to what to sell, developing new brands, packaging and labeling of its own brand, and customer service policies (a vital extension of a product).

One of the pillars in an merchandising strategy is to offer consumers the largest selection of products at each store. This allows the customer to make an informed choice as to what to purchase.

Large supermarket chains, like Safeway, also offers a wide range of high quality brand name products produced at its manufacturing facilities. Lucerne Dairy products, Mrs. Wright bakery items and TownHouse grocery products are examples of just a few. New products are constantly developed to meet the needs of the changing requirements of customers.

Satisfaction guaranteed is Safeway's number one customer service policy. If the product does not satisfy a customer's needs, Safeway will gladly provide an exchange or refund- no questions asked!

Pricing

Competitive pricing is another pillar of a merchandising strategy. Supermarkets strive to offer consumers the ultimate shopping experience by providing total value. Total value is defined as a combination of superior quality, selection and service at competitive prices—all provided in attractive, well-maintained facilities.

Distribution

The term "distribution" is used to refer to the channels through which a product is sold in the market place. Often, a distinction is made between wholesaling and retailing channels. Wholesaling refers to those functions of gathering products, breaking lots into smaller units and distributing to retailers. Retailing refers to the location where the end customer can directly purchase the product.

For example, Safeway Inc. with 1,100 retail stores in Canada and the United States, is one of the largest grocery retailers in the world. These stores carry products from thousands of food and consumer goods manufacturers, including Safeway brand products produced at its 56 manufacturing locations.

Promotion

Every business needs a means of communicating with its customers (and potential customers) to promote its products or company image. Promotional activities tend to fall into two categories: strategies that "pull" the product toward the end user by enticing the consumer to buy the product, and "push strategies" strategies that force the product through the entire product distribution channel by focusing on salespeople to sell the product.

The best known promotional pull strategy is advertising. Advertising is geared to attract a customer's attention to help win his/her loyalty. Radio, television, billboards and the printed press carry the bulk of the advertising messages. Most companies out-source (hire outside companies) to develop their advertising campaigns. Consumer sales promotions such as coupons, double coupons, free samples, contests and point-of-purchase displays are aimed at convincing a customer to "try" the product at a relatively low risk.

Push strategies include personal selling, sales incentives and sales promotions geared at members of the distribution channel to stock the product and sell it to the next "buyer".

For example, Safeway will spend an estimated $35 million on a current advertising campaign aimed at promoting the company and its products in the market place.

Although companies such as Safeway may have an overall "standardised" promotional strategy, individual retailers or stores may focus on communicating with particular consumer groups within their target area. For example, a promotional campaign might be geared at elderly customers in neighborhoods with large concentrations of older adults. Such a group of customers with specific requirements and market potential is referred to as a market "niche."

Customers

No discussion of customer service would be complete without a definition of a customer. Like grocery stores, high schools are business enterprises. They have multiple customers. A customer is anyone, internal or external, who is affected by the processes or services of a business enterprise. The various customers of a high school are the students, the parents, the teachers, the Board of Education...as well as the future recipients of the products of high school education, such as colleges, universities and future employers.

Customer Chain

Each one of these groups is part of the broader customer chain, some as internal customers, others as external customers. An external customer is someone who directly purchases the products or services of a business enterprise. An internal customer is a group in the distribution channel that receives processes, products and services from others in the organisation. For example, your teachers are customers of one another. If you fail to become a proficient reader or writer after taking English classes, you will not be an effective student in a history class. In this case, the history teacher is an internal customer of the English teacher.

The Quality Paradigm

A revolution is sweeping across all industries, including the grocery business. To provide superior customer service, businesses are changing the way they look at all facets of an operation. We sometimes refer to a revolution or change in thinking as a "paradigm shift." Scientist Thomas Kuhn theorised that a revolution occurs not because of startling new facts, but because of a change in the overall way something is seen.

Thus, the paradigm shift that is taking place in industry is not necessarily due to a gain in knowledge, but a change in the way customer service is viewed. The term "quality paradigm" refers to a new way of thinking about quality and customers. Nothing in a business enterprise can be taken for granted. Each work process must continuously be improved in order to remain competitive and attain world-class status. The quality paradigm is the foundation of the entire Total Quality Management (TQM) revolution in industry.

Total Quality Management

Before a company can provide superior customer service, it must have an overall strategy. TQM is a management philosophy geared toward the continuous improvement of quality to meet, exceed and anticipate customer expectations. TQM focuses on all customers, both internal and external.

The requirements to TQM:

— A long-term company commitment

— A change in employee attitudes and behaviors

— A collaborative team effort among the various functional departments

— An emphasis on the empowerment and involvement of all employees

Customer Satisfaction

Customer satisfaction and loyalty represent a company's long-term financial success. Without first understanding the reasons for lack of customer loyalty, no amount of planning will lead to success. Here are some facts:

— A typical supermarket customer in the Northern California region spends an average of $65 a week on groceries, so the value of that customer over a I0-year period (adjusted for inflation) is $40,600.

— Research on grocery shoppers (Progressive Grocer) has shown that 69% stop shopping at a particular store because they are dissatisfied with its service or employee attitudes. Rather than discussing the problem with a store manager or clerk, 98% of dissatisfied customers simply go to a competitor's store.

— A customer who experiences a "moment of truth," or an unpleasant incident, will tell five to seven other people about it. This negative word of mouth is obviously very damaging to a company's reputation and goodwill.

Considering these facts, it is of utmost importance to continuously improve customer service. The success of a company is contingent on measuring customer satisfaction, developing action plans to close service gaps, and redesigning the work process to streamline operations.

Commitment of quality and customer satisfaction

Supermarkets emphasise superior service as one of their key competitive strategies. Everything a supermarket employee does affects other employees (internal customers)

as well as shoppers (external customers). In addition to continuously improving their technical expertise, employees focus on customer relations and satisfaction. Achieving superior service is the most important job of everyone—whether in a retail store or other parts of the organisation, such as manufacturing, distribution, warehousing, finance or human resources.

As an example, Safeway's service culture targets four pillars of success in the 90's. These pillars are price, service, quality and selection. Sustaining this service orientation among the 110,000 employees is a long-term effort that involves the commitment and dedication of everyone.

According to the TQM philosophy, everyone has customers. Therefore, each job has a customer service component. There are a number of professional careers dealing directly with quality customer satisfaction. They include quality control supervisors and managers, TQM team leaders and members, quality facilitators and trainers, customer satisfaction researchers, human resource managers, directors of quality, etc.

To work in the area of quality and customer satisfaction, you must possess good interactive, facilitative and teamwork skills as well as a sound statistical and analytical background. More and more colleges and universities are integrating TQM elements in their professional education programs. Business, engineering and food science schools are starting to emphasise, usually on an elective basis, the need for continuous quality improvement subject matters. Some universities have TQM as specialty area of concentration.

Food Preparation at Home

Freezing

Freezing can be thought of as a continuation of

refrigeration. If we lower the food's temperature beyond a certain point, it will freeze. The freezing point for pure water is 0°C. For food, however, the freezing point is below that. You may recall from physics that the presence of solutes in water will lower its freezing point. Water contained in foods has many solutes, mainly salts and sugars.

Freezing is similar to refrigeration in that it will not destroy microorganisms or inactivate enzymes, but rather slow their deteriorative effect. In some cases, it will completely inhibit the activity of microorganisms, even though they will still be alive. Enzymes, on the other hand, will maintain a certain level of activity during freezing. This is the reason we sometimes blanch vegetables prior to freezing them. Blanching is a mild heat treatment designed to inactivate enzymes.

Freezing has been a key technology in bringing convenience foods to homes and restaurants. It causes minimal changes in the quality of food in terms of size, shape, texture, color, flavor and microbial load. This is assuming that the freezing process is carried out properly. A key factor during food freezing is how fast we freeze. It is known that when food is frozen slowly, the ice crystals formed will be large. Large ice crystals are undesirable as they will damage the cell structure of foods like meat and fish. This will result in poor texture and excessive dripping upon thawing. It is therefore important that freezing be carried out at a very fast rate. As is the case with refrigeration, proper cleaning and packaging prior to freezing will yield high quality frozen food.

Frozen food must be maintained at the proper temperature at all times. Ice cream, for example, is very sensitive to fluctuations in temperature that can cause a defect called sandy texture. The ice cream will actually taste grainy, as if it contained sand granules. These granules are in fact lactose crystals. Lactose is the naturally occurring sugar in milk that is normally dissolved. Temperature fluctuation, however, will cause lactose to crystallise out, imparting sandiness.

The success of freezing technology has opened a new field for food processors. They can prepare complete meals and freeze them until the consumer is ready to thaw and heat them. Many of these meals are sold in their serving dishes. Other popular frozen foods are pot pies, fish sticks, desserts and potatoes. At present, there is no other form of food preservation that offers the convenience of frozen foods.

Food processors freeze foods to an internal temperature of -18°C. The food must be maintained at this temperature or slightly lower during transport and storage. Many fruits and vegetables will retain good quality at the above temperature for up to 12 months or even longer. The expected frozen storage will vary with temperature. For example, frozen orange juice will last 27 months at -18°C, 10 months at - 12°C, and only 4 months at -6.7°C.

The popularity of microwave technology has contributed to an increase in frozen food consumption. Both developments in food processing and preparation enable us to have convenient, tasty and nutritious meals at a reasonable price.

Refrigeration

Refrigeration or cold storage of food is a gentle method of food preservation. It has minimum adverse effects on the taste, texture, and the nutritional value of foods. It must be kept in mind, however, that refrigeration has a limited contribution towards preserving food. For most foods, we can expect refrigeration to extend the shelf-life by a few days. In many cases, refrigeration is not the sole means of preserving the food. For example, milk is first heat processed (Pasteurised), and then refrigerated.

The heat treatment destroys the pathogenic microorganisms and reduces the total microbial load as well as inactivates the enzymes in the milk. Refrigeration will keep the spoilage reactions (microbial or enzymatic) to a minimum. Remember, refrigeration does not kill microorganisms or inactivate enzymes, rather it slows down their deteriorative effects. Refrigeration temperature is a key factor in predicting the length of the storage period. For example, meat will last 6-10 days at 0° C, one day at 22° C and less than one day at 38° C. Household refrigerators are usually run at 4.7 -7°C.

Commercial refrigerators are operated at a slightly lower temperature. On an average day, the room temperature is 25° C, so the longer you leave food on the kitchen counter, the sooner it will spoil. Most spoilage microorganisms prefer warmer temperatures, but there are a group of microorganisms called psychrophilic which will grow at refrigerated temperatures. When cooling is used as a preserving technique, it is essential that the food is maintained at the proper cold temperature during manufacturing, transport, display and home storage.

Signs of spoilage in refrigerated foods vary with the food product. In fruits and vegetables loss of firmness or crispness takes place. Red meat will change in color, while fish will get softer with noticeable drippage. Refrigerated food will last longer if it is cleaned and properly packaged prior to refrigeration and, of course, maintained at the proper temperature with minimum exposure to surrounding temperatures.

Since refrigeration is an energy-demanding process that must be maintained throughout the life of the product, refrigerated foods tend to cost more than non-refrigerated foods. Yet, today's consumer is buying more refrigerated foods for their fresh quality. When attempting to preserve the high quality of refrigerated foods, the consumer need only follow one key rule: keep cold food cold.

Microwave

Four out of five American homes have at least one microwave oven. Because they heat food very rapidly, microwave ovens are extremely popular. A microwave oven is a device that generates microwaves from electricity. Microwaves are electromagnetic energy that have an electric and a magnetic component. Other forms of electromagnetic energy are radio waves, sunlight and electricity.

Scientists refer to microwave energy as radiation because, similar to other radiation energy, it travels through space. Microwave radiation is often called non-ionizing radiation to distinguish it from other forms of radiation, like x-rays, which are called ionizing radiation. Ionization refers to breaking atoms or molecules into two electrically charged groups.

Microwave energy, like all electromagnetic radiation, travels in a wave pattern. The waves are reflected by metals, pass through air, glass, paper and plastic, and are absorbed by food. Most microwave containers are designed to transmit microwave energy without reflecting or absorbing it, and thus are made of paper or plastic. The microwaves will travel through the container to the food.

When food is exposed to microwaves, it will absorb that energy and convert it to heat. Food composition (mainly water content) is a key factor that determines how fast it will heat in a microwave environment. The higher the water content of a food, the higher its loss factor thus, the faster it will heat. Solutes, like sugar and salt, also influence the loss factor of foods exposed to microwave energy. For foods to heat in a microwave oven, the electromagnetic waves must penetrate the food. There are limits to the depth of their penetration. This makes container geometry a very important factor in heating foods by microwave energy.

Food composition does not only influence the loss factor, but also penetration depth. For example, when salt is added to water, it will change its microwave heating characteristics in two different directions. On one hand, it will increase the water's loss factor causing the water to heat faster. On the other hand, salt will decrease the penetration depth of microwaves into the salt/water solution, decreasing the heating rate. If salt is added to water to enhance its loss factor and, at the same time, the container geometry changes to minimise the drop in penetration depth, the heating rate will be greatly increased.

There are increasing numbers of food scientists who are dedicated to developing microwavable foods that can heat quickly and evenly while maintaining high quality. Heating foods evenly in a microwave oven is difficult at best, particularly with solid foods of different composition. A frozen dinner tray is an example of a solid food with varying composition. Have you ever tried to warm a jelly-filled donut in a microwave oven? If you have, you would have noticed that the filling gets extremely hot while the cake part is hardly warm. This is because the jelly and the cake parts both have different loss factors and therefore will absorb the microwave energy at variable rates and heat to totally different temperatures. A simple way to avoid this problem is to let the food stand for a short period with the microwave oven off. This will permit heat to travel from the hot regions of the food to the cooler ones, through a process called conduction. Eventually, the food will equalise at a temperature somewhere between the hot and the cold regions.

Understanding how food composition influences microwave heating is a skill needed by those developing microwavable foods. Developing new microwavable meals and snacks is a growing segment of the food industry. Aside from using microwave energy to heat meals at home, there have been many attempts to utilise microwave energy to heat food at processing plants. Microwave energy is also utilised in the textile and paper industries to remove moisture from the products.

2

Production of Food Products

Food is traditionally obtained through farming, ranching, and fishing, with hunting, foraging and other methods of subsistence locally important for some populations, but minor for others. In the modern era, in developed nations, food supply is increasingly dependent upon agriculture, industrial farming, aquaculture and fish farming techniques which aim to maximise the amount of food produced, whilst minimising the cost. These include a reliance on mechanised tools which have been developed, from the threshing machine, seed drill, through to the tractor and combine, etc. These have been combined with the use of pesticides to promote high crop yields and combat those insects or mammals which reduce yield.

More recently, there has been a growing trend towards more Sustainable agricultural practices. This approach - which is partly fuelled by consumer demand - encourages biodiversity, local self-reliance and Organic farming methods.

Major influences on food production are international policy, *e.g. the World Trade Organization and Common Agricultural Policy*, national government policy or law and war. Food for livestock is fodder and traditionally comprises hay or grain.

Animal Cloning and the Production of Food Products

For the first time in history, it is commercially feasible to clone animals for use in food production. New technology, known as somatic cell nuclear transfer (SCNT), allows the production of large numbers of clones with a high degree of control over many of their traits, including gender. The cloning of Dolly the sheep first brought SCNT to the public's attention in 1996. While societal concern focused on the prospect of human cloning, the agriculture and biotechnology industries quickly realized the potential of cloning for food production and medical uses. It was not long before cattle, goats, pigs, mice and rabbits had all been cloned. At the same time, the consuming public may be ill-prepared for the arrival of food products from cloned animals on their grocery shelves.

Concerns range from the moral and ethical implications of cloning to the environmental safety of cloned animals. As a society, we have yet to have a well-informed and thoughtful public discourse on animal cloning. Even among food producers and retailers, cloning may be seen variously as the next natural step in animal breeding techniques, or a radical change affecting livestock production and marketing.

Today, innovative companies are ready to sell food products from cloned steers and dairy cows. They are observing a voluntary moratorium on the sale of their products, while waiting for guidance on marketing cloned products from the U.S. Food and Drug Administration (FDA). The FDA's Center for Veterinary Medicine has been charged with evaluating both the food safety and animal health implications of cloning, as well as its environmental impact.

Dr. Steven Stice, of Prolinia, Inc., a company focused on cloning animals for agricultural production, began by distinguishing cloning from the development of transgenic animals. Although the two technologies can be complimentary, "by no means should anybody think that cloning is genetic engineering," he said. The distinction between the two lies in that cloning is essentially asexual reproduction-it produces offspring that are genetically identical to the single parent organism, whereas transgenic technology introduces genetic material from one organism into another.

Cloning animals, however, is far more recent and complicated. The earliest form of animal cloning, begun in 1978, involved splitting an already formed embryo. A second technique, called blastomere nuclear transfer (BNT) fuses the nucleus of an embryo cell with an unfertilized egg that has had its own nucleus removed. Both techniques, however, afford little control over the cloned animal's traits, even the most basic characteristic of sex, explained Stice. Therefore, he observed that if the split embryo of a prized milk cow yielded two steers, the effort was wasted and the animals often sacrificed. These early techniques were further limited in that they yielded few animals per embryo.

Somatic cell nuclear transfer (SCNT), however, overcomes these limits, explained Stice. SCNT allows breeders to copy an adult animal whose traits are well-known. Biologists can work with thousands of eggs removed from an animal's ovary at one time and thousands of nuclei taken from cell lines cultivated in the laboratory, and thus conceivably produce thousands of clones from a single animal.

In 1996, researchers performed the first successful adult or somatic cell nuclear transfer (SCNT) cloning with a sheep they named Dolly. Stice pointed out that although Dolly was the public's first major exposure to animal cloning, researchers had been involved in cloning efforts for more than 20 years. "It hasn't been all overnight," he said. The new technique fuses the nucleus from a differentiated adult animal cell (for instance, a skin cell or kidney cell) with an unfertilized egg that has had its own nucleus removed. The new nucleus contains all the genetic material needed to create the clone. However, biologists must reprogram the reconstituted egg to behave as if it has just been fertilized, and thus initiate embryo development. Finally, the embryo is artificially inseminated into a surrogate mother where, if all goes well, it will gestate until birth.

Many of the problems with cloning are attributed to "epigenetic effects," in which identical DNA can produce different outcomes. The consequences can be as harmless as unique color patterns in the coat of a cloned cow, or as damaging as placental failure and cardiovascular birth defects. A number of factors cause epigenetic effects. These include the interaction of the donor nucleus and the unfertilized egg; environmental factors, such as nutrition, available in utero; and in vitro culture conditions.

That epigenetic effects are not limited to clones. "Anything from in vitro fertilization to artificial insemination may have some epigenetic effects," he said. Even identical twins produced from conventional mating experience epigenetic effects, as seen in their different fingerprint patterns and different iris pigmentation patterns. The goals of cloning as moving better genetics to the market, preserving genetic diversity, and creating more uniformity in herds, which could benefit both the environment and the producer.

Dr. Eric Hallerman is a professor at Virginia Polytechnic Institute and State University in US, and member of the committee that produced the National Research Council's report, Animal Biotechnology: Science-Based Concerns. Hallerman reviewed the report's key findings regarding animal cloning. He focused on issues related to the food safety of cloned products and cloning's impact on animal welfare, as well as public policy concerns. Hallerman began by reviewing the main scientific issue underlying food safety concerns. He described this as, "whether and to what degree genomic reprogramming occurs" in the process of making a clone.

"The offspring formed will be identical to the donor in terms of the nuclear DNA, but will differ in terms of mitochondrial genes [found in the egg cytoplasm], and possibly also in the manner that genes are expressed and biochemically modified," he said. Hallerman said that there is little scientific information available regarding the impact of such reprogramming on food safety, either in BNT or SCNT clones. "Although several studies are in progress, there are to date no published data comparing the composition of meat and milk products of somatic cell clones, their offspring, and conventionally bred animals," Hallerman said.

However, he pointed out that BNT clones have been on the market for decades with no reported ill effects, and that there is no evidence to date that SCNT clones pose a food safety risk. Hallerman also reviewed the committee's findings on animal welfare issues. He said that it considered several aspects of animal welfare related to cloning, including the potential to cause pain, physical or psychological distress, behavioral abnormality, physiological abnormality or health problems, and the technology's potential to alleviate or reduce these problems.

These technologies, which are also commonly used in commercial livestock breeding, include super-ovulation, semen collection, artificial insemination, embryo collection and embryo transfer use. Some of these may cause pain or distress, or require that the animal be killed to harvest eggs or semen. In particular, Hallerman noted, "Ruminants produced by in vitro culture methods tend to have higher birth weights and longer gestation times than calves or lambs produced by artificial insemination, a phenomenon referred to as large offspring syndrome (LOS)."

LOS offspring have more congenital malformations and higher perinatal mortality rates, explained Hallerman. Furthermore, because of LOS, calving can be difficult, leading to more cesarean sections. For example, "Thirteen percent of in vitro produced calves are delivered by emergency cesarean section, as opposed to less than one percent of calves produced by standard artificial insemination," he said.

Hallerman then examined animal welfare issues specific to SCNT technology. Cloned cattle and sheep, commonly have problems with high prenatal and perinatal mortality, abnormal placentas, developmental abnormalities, and LOS. While not seen universally among cloned animals, some researchers have reported respiratory distress, lethargy, lack of suckling reflex and a number of metabolic problems in the immediate postnatal period, he stated. "It's difficult to determine which problems are due to cloning by nuclear transfer, per se, which are due to embryo culture and transfer methods or to some interaction of those different technologies," he said.

Hallerman also discussed concerns the committee had on cloning and animal behavior. He explained that researchers had observed behavioral impairment of SCNT cloned mice, though the impairment appeared to be transient. The committee had therefore recommended studies on the behavior of cloned animals. Since cloning by its nature produces identical copies of the donor individual, it reduces genetic variability relative to reproduction through conventional breeding. Such a loss could limit the potential for future genetic improvement, Hallerman said.

In addition, disease could spread more rapidly through a genetically uniform herd. On the other hand, he pointed out that cloning provides the opportunity to preserve genetic variability by allowing scientists to use genetics from animals that either had reproductive problems, or that had died, but from which tissues had been preserved. Hallerman made clear that SCNT was not an efficient process for reproduction, given that less than 10 percent of the embryo transfers result in live offspring. However, he said that the success rate appears to be increasing, and animals that survive to adulthood appear to be normal.

Cloned Food Products

Milk from cloned cows and meat from the offspring of cloned cows and pigs could show up on grocery shelves as early as next year under the plans of livestock breeders who are already raising scores of clones on American farmsteads.

A recent report from the National Academy of Sciences, the nation's top scientific body, has given fresh impetus to the effort to turn cloning into a routine tool of U.S. agricultural production. A special NAS panel that reviewed developments in animal biotechnology was alarmed by genetic manipulation of fish and insects that might escape and harm wild species, but it found cloning of farm animals less troublesome, since the technique involves copying adult animals without altering their genes. The committee called for a few additional studies but said the technique was unlikely to affect the safety of the food supply.

A few cloned cows scattered around the country are already producing milk. Farmers and companies have held off selling it only because of informal requests from the Food and Drug Administration, which is reviewing whether clones, their byproducts or their offspring should be allowed into the food supply. The agency hopes for a decision by late this year. Absent compelling evidence of a problem, it's not clear the FDA or any other government agency would have the legal power to keep cloned animals out of the food supply.

The biggest lingering concern appears to be for the welfare of cloned animals and their surrogate mothers, with some groups saying the development of cloning on a large scale will lead to widespread animal suffering. The number of cloned animals living on American farms today is small — most estimates put it at fewer than 100. All are elite animals that cost tens of thousands of dollars to produce and are valued as breeding stock, not as meat.

For that reason, products made from clones are likely to be a mere trickle in the marketplace, at least at first, farmers said. But if the technique becomes established and the price of cloning falls, farmers envision a day when entire dairy herds may be stocked with nothing but clones of the most prolific cows.

The technique may never be cheap enough to produce animals for use directly as meat, but breeders said it's likely the first- or second-generation offspring of clones will wind up in the meat supply in large numbers. The cloned food products that could hit the market in small quantities next year include milk from cloned Holstein dairy cows and, potentially, veal from their first-generation offspring. Pigs would likely not be far behind, with some first-generation offspring probably being butchered for food in 2004 or 2005, animal breeders said.

Other countries are moving in the same direction as the United States. A study published recently in Japan said cloned meat and milk are identical to the ordinary kind. That nation is now preparing to lift a ban. A handful of American animal breeders plunged into cloning two years ago, mindful that no law or regulation said they couldn't, and they were caught off guard by the informal FDA moratorium that followed. They said they had been trying to honor the agency's ban, but economic pressures are mounting as the animals reach their prime reproductive years and they need to recoup investments that typically exceed $20,000 per clone.

Some cloning companies are already pouring out milk from clones. By this spring, farmers in scattered spots around the country will face the choice of selling milk from their clones or dumping it back on their fields. If the FDA will let them, they plan to sell it. Greg Wiles, proprietor of a dairy farm in Williamsport, Md., knows the pressures firsthand. His family once owned the top-ranked Holstein in the country — Con Acres HS Zita-ET, registered Holstein No. 14411844. Zita, worth as much as $150,000 in her prime, died last year — but not before Wiles, with help from a company in Massachusetts, spent some $70,000 to create two clones, Genesis and Cyagra.

They are healthy animals who have reached sexual maturity, Wiles said, and he has already used embryos from the clones to produce seven new pregnancies by embryo transfer to recipient cows. Wiles said he needs to establish pregnancies soon in the clones themselves. That will not only let them build milk-production records, it will also help them keep producing embryos at a high level in their mature years. The clones and their female offspring, if they are anything like Zita herself, should be fantastic milk producers, Wiles said.

"We are probably getting a little ahead of the FDA," Wiles said. "I have always stood behind the fact that we knew we weren't genetically altering anything with the cloning process. There was a lot of newness when the clones were born, but they've really become just like any other cow on the farm."

Even breeders who have not yet embraced cloning are preparing for the day it becomes routine, paying several hundred dollars to freeze cells from elite animals that could eventually be used to make clones. Some have learned the technology can salvage the genes of some animals that have already died.

Jon Fisher's business is selling semen from champion boars to breeders who produce the long, svelte pigs that dominate 4-H and Future Farmers of America livestock competitions. The new animal, much in demand, greatly elevated Fisher's reputation in the world of pig breeders. He named the boar 401-K, after the retirement account. A year ago Fisher was getting ready to have 401-K cloned when the animal died suddenly of an intestinal blockage. The boar had been dead several hours by the time Fisher managed to salvage ear cells and ship them off to Infigen Inc. of DeForest, Wis., one of a handful of American companies offering cloning services to breeders. "It was like a bad Woody Allen movie, the way we were running around here," Fisher said.

3

Community Food Systems

The food system includes all processes involved in keeping us fed: growing, harvesting, processing (or transforming or changing), packaging, transporting, marketing, consuming and disposing of food and food packages. It also includes the inputs needed and outputs generated at each step. The food system operates within and is influenced by social, political, economic and natural environments. Each step is also dependent on human resources that provide labor, research and education.

Several qualifying terms have been used to describe the food system: simple, complex, local, global and regional. A community food system is a food system in which food production, processing, distribution and consumption are integrated to enhance the environmental, economic, social and nutritional health of a particular place. A community food system can refer to a relatively small area, such as a neighborhood, or progressively larger areas - towns, cities, counties, regions, or bioregions.

The concept of community food systems is sometimes used interchangeably with "local" or "regional" food systems, but by including the word "community" there is an emphasis on strengthening existing (or developing new) relationships between all components of the food system. This reflects a prescriptive approach to building a food system, one that holds sustainability - economic, environmental and social - as a long-term goal toward which a community strives.

Four aspects distinguish community food systems from the globalized food system that typifies the source of most food we eat:

— food security,

— proximity,

— self-reliance and

— sustainability.

Food security is a key goal of community food systems. While food security traditionally focuses on individual and household food needs, community food security addresses food access within a community context, especially for low-income households. It has a simultaneous goal of developing local food systems.

Proximity refers to the distance between various components of the food system. In community food systems such distances are generally shorter than those in the dominant or global food system. This proximity increases the likelihood that enduring relationships will form between different stakeholders in the food system - farmers, processors, retailers, restaurateurs, consumers, etc.

Self-reliance refers to the degree to which a community meets its own food needs. While the aim of community food systems is not total selfsufficiency (where all food is produced, processed, marketed and consumed within a defined boundary), increasing the degree of selfreliance for food, to be determined by a community partnership, is an important aspect of a community food system.

Sustainability refers to following agricultural and food system practices that do not compromise the ability of future generations to meet their food needs. Sustainability includes environmental protection, profitability, ethical treatment of food system workers, and community development. Sustainability of the food and agriculture system is increased when a diversified agriculture exists near strong and thriving markets, when non-renewable inputs required for every step in the food system are reduced, when farming systems rely less on agri-chemical fertilization and pest control, and when citizen participation in food system decision-making is enhanced.

Goals of Community Food Systems

Building a community food system requires comprehensive or holistic approaches to meeting the food needs of people living in a particular place. Efforts to develop community food systems address multiple goals simultaneously:

— Optimized health, reduced risk of diet-related chronic diseases, and increased enjoyment of food among community members.

— Dietary change that complements the seasonal availability of foods produced and processed by the local food and agriculture system.

— Improved access for all community members to an adequate, affordable, nutritious diet.

— A stable (or in some cases, expanding) base of family farms that use integrated production practices to enhance environmental quality,

— Marketing channels and processing facilities that create more direct links between farmers and consumers, and, by shortening the distance between these partners, conserve resources needed for transporting food.

— Food and agriculture-related businesses, resulting in stronger community economies through job creation, and re-circulating financial capital in the community. Such businesses could include food processing, or value-adding processing to expand opportunities for locally produced food to be consumed locally.

— Increased public participation in food and agriculture policies that promote local food production, access to local retail and processing markets, and institutional procurement of local agricultural commodities.

Elements of Community Food Systems

There are several well-recognized elements of a community food system:

Farmers' markets provide the opportunity for eaters to meet and talk directly with the people who grow the food they are buying. By the same token, farmers can learn more, in a direct way, about what their customers want and need to know about the food from their farms. By decreasing the amount of fuel used to move food around, this proximity to food sources increases the environmental sustainability of the system.

Community and school gardens are recognized as an important source of fresh produce, particularly for underserved populations in low-income neighborhoods, thereby increasing dietary quality and food security. They provide spaces for community interaction, decision-making, problem-solving, creativity and celebration. Community gardens also provide opportunities to learn about food production, develop job skills, increase agriculture literacy, generate food-related businesses, and create links to nearby restaurants and soup kitchens.

Community supported agriculture (CSA) farms are arrangements whereby a group of people buy shares into the eventual harvest of a farm before the crops are planted. In exchange for their investment into the farm, shareholders receive fresh fruits and vegetables (and sometimes, other products such as local cheeses, fresh flowers, eggs and meats), on a weekly basis throughout the harvest season. By making this investment, CSA members accept part of the financial risks associated with farming. Further, the farmer receives a portion of the cost of production at a time when it is most needed. Many CSA farmers also market through local farmers' markets, which can increase farm profitability and stability.

U-Pick operations and roadside farm stands provide access to fresh produce direct from the farmer who grew it. Through a U-Pick, the price paid to the farmer is reduced substantially in exchange for harvest labor. In the process, eaters come in contact with farmers, experience another aspect of the food system, and increase their intake of fresh and processed local produce (if some of what they harvest is preserved through freezing or canning, for example).

Community kitchens are facilities where locally produced, gleaned or recovered foods can be further processed or preserved for members of a community. Food product development often takes place at these facilities, thereby creating income generating opportunities and products with local identity.

Small-scale food processing and decentralized root cellars provide infrastructure and technical expertise necessary to launch new foodbased businesses. Much of the food we eat is processed in some way and in areas with relatively short growing seasons, such as the Northeast. The use of canned, frozen and stored fruits and vegetables when produce is "out of season" is another way to develop community food systems.

Externalities

The word externality is an economic term used to describe costs or benefits generated by an agent (say a farmer, or a truck driver) that do not register as a cost or benefit to that agent or end-user. The pollution generated by transporting food is not paid for by the trucking company in the price of the fuel, or by the consumer in the price of the food. The external environmental and social costs related to food production, processing, storage, and distribution are seldom accounted for in the price we pay for food at the grocery store register.

Community food systems, by narrowing the distance between producers, processors and consumers, have a greater chance of "internalizing" any externalities in the food system and actually reducing many. For example, since the distance food is transported in a community food system is shorter, less fossil fuel is burned, less pollution generated and less wear and tear on trucks and roadways results from the transportation of food. Likewise, because more of the steps in the food system are carried out locally, the loss of food system-related jobs is minimized.

Creating a Community Food Systems

As individuals, consumers can do a lot to support and collectively strengthen community food systems:

— choose a diet rich in locally grown and processed foods. Regional food guides, provide guidelines to help consumers choose healthful local and seasonal diets.

- ask food stores to buy from local growers and processors.
- ask where items on restaurant menus came from and express interest in eating locally produced and processed foods.
- shop at farmers' markets and food co-ops (which are more likely to offer local, in season, and often organic choices).
- buy a share in a CSA farm or sponsor someone else's share.
- participate in a community or school garden or start a home vegetable garden and share excess with neighbors, a community kitchen or local soup kitchen.
- cook from scratch.
- support policies that favor local farms and other elements of community food systems, join or create a food policy council to assess community assets with respect to the local food system, identify areas of need, and develop strategies collectively to meet those needs. In order to support local community food systems in their food choices, consumers need:
- access to local foods,
- ways to identify local alternatives,
- ways to learn meal planning and preparation skills,
- an understanding of seasonal variation,
- knowledge of the local food and agriculture system, and
- an appreciation of the benefits of eating seasonally and locally.

Nutrition practitioners can do a lot through their professions to support community food systems as well, such as:

— include considerations about seasonal availability of locally produced foods when providing dietary advice to clients,

— substitute non-local foods in meal plans with foods that are nutritionally equivalent and are produced locally,

— create seasonally varied institutional food service menus that reflect local agricultural production. This might include, for example, the use of root vegetables in the winter in northern climates,

— shift procurement strategies in food service operations toward local food sources, and include information about the sources of foods at the point of purchase.

4

Electronic Systems in Food Industry

Before electronic information technology was available, the food supply chain was segmented in several separate business sectors or silos. Farm input supply businesses, farm producers, and the first line handlers and shippers make up the agricultural production system. The next sector in the supply chain is made up of first line processors, referred to as the agribusiness sector. They process raw commodities into ingredients that generally need further processing or cooking to make edible food for human consumption, or they package fresh produce that is sold directly into stores. Next are food manufacturers who turn ingredients into finished food products that typically are sold to wholesale warehouses or retailer owned distribution centers.

Finally, there is the retail sector made up of retail food stores and consumers. Since the raw material to manufacture food is seasonal, inventories tend to build up in the manufacturers' warehouses. They offer wholesalers discounts to "forward buy." If the wholesalers' inventory builds up they, in turn, offer retailers a discount and other services to "forward buy," pushing the costs of holding inventory farther down the food chain. This system, know as a supply push system, relies heavily on well know manufacturers' brands and national advertising to move products. It worked well for wholesalers to hold excess inventory in an era of high inflation because they could buy at a discount and sell at inflated prices before the shelf life of the product expired. With low inflation and rising costs of space, labor and management, bulky inventories, like the elephant in the python, became a financial burden.

Traditionally, manufacturers and retailers have been adversaries, each trying to extract the maximum profit margin from the final sale price of the product. Wholesalers and retailers often charged, or manufacturers offered, (slotting) fees to introduce and handle new products. Retailers relied on manufacturers' advertising, couponing and promotion campaigns to help sell the products in their stores. Retailers, whose business territory is considered to be three to five miles around their store, competed with each other on price and consumer services. They were secretive about their merchandising strategies and operations.

Competition within the industry for a share of consumers' food dollar (the share of their stomach) is fierce. In the twelve-year period between 1987 and 1999 inflation adjusted sales in eating and drinking places grew an average of 2.2 percent while similar sales in retail food stores decreased an average of 0.1 percent. Consumers can obtain food everywhere these days, including vending machines, drive through places and the Internet. Retailers soon realised that their biggest competitors were not other stores but fast food or quick service restaurants. This has been an industry dominated be many small independent businesses at both ends. There were many farmers, many retailers, and relatively few manufacturers and distributors.

Farmers and retailers each felt that the larger companies in the middle were profiting at their expense.

Farmers' defense was to organise into buying or selling cooperatives and/or lobby for government price supports and access to foreign markets. Retailers' strategy was to buy low and sell as high as possible consistent with growing customer sales in a highly competitive sector. There was little vertical integration or organisation. Perhaps this is because the businesses at each end of the food chain need to be widely dispersed across the landscape.

Agriculture production is tied to the appropriate climate and land bases and grocery stores have to be located in each village where people live. Whatever the reason, many independent business operators valued their independence and believed in their value to society. Into this fragmented and fiercely independent system came a single fierce competitor called Wal-Mart. It was able to lower retail prices by developing an integrated supply/demand chain driven by the sharing of information about retail sales with suppliers in real time.

Electronic technology made the collection, analysis and transmission of data possible, but Wal-Mart developed and perfected the system. It is fair to say that they forced the rest of the industry to adopt e-commerce for business practices and to build new relationships with their suppliers. Ironically, food retailers owned, and ignored, a key resource for improved efficiency in this supply chain since the mid 1970's. The data scanned every day at their check out counters is the beginning of the information chain for business to business e-commerce relationships. Now they have learned to use that data.

Business to Consumer E-commerce

Internet ordering of food is a modern version of an old practice in retail food, a practice that was abandon because it was too expensive. In the first half of the twentieth century, small town or neighborhood grocery stores carried customers' credit accounts, took phone orders and delivered food to their homes. But, with product proliferation, consumers needed to see the new products in order to make choices. With automobiles consumers became mobile and it was less expensive for them to stop and shop as they traveled around the suburbs for other purposes.

Consumers used their own time to provided "free labor" for shopping and delivering groceries to households. Groceries became cash and carry stores, then suburban supermarkets, and now supermarket chains and super centers where customers bag and haul their own groceries. With the Internet, customers can once again purchase on credit and have food delivered to their homes. Alternative models of shopping include phone and fax orders to grocery stores that might be delivered or simply bagged for the consumer to pick up.

What has changed to make home delivered groceries attractive once again? The advent of "time starved" consumers and their access to the Internet makes home delivery look like a solution to a modern consumer problem. Most surveys show that consumers do not like grocery shopping, considering it a drudgery task. This type of shopping is ripe for Internet competition. Weekend or occasional shopping is leisure. It is entertainment, fun, an adventure and a social event. The Internet cannot compete with this activity by selling and delivering product to the household's doorstep.

Internet shopping for food represents enormous opportunities to take product and service to consumers in a most convenient way. It is estimated to grow to $6.3 billion in 2000 and $11 billion by 2003 (which is still only 2.3 percent of 1999 grocery sales). However, finding a way to lower the costs of shopping (picking) and delivering that competes with the free labor provided by consumers is a challenge. At least one-third of Americans is "on line" today. Of those who are on line, 77 percent believe that going on line makes their lives better. Consumers are finding that the online services are convenient. Ordering and shopping (browsing) any time of the day or night, savings on transportation to a mall, and receiving products in one's home or office is convenient, and convenience is a primary quest for time pressed American consumers.

Features of Business to Consumer E-commerce

The advent of e-commerce for home shopping increased competition for some traditional retail stores and offered a new form of business to others. Initially, Internet sellers partnered with bricks-and-mortar retailers and used their stores or distribution centers as a source of the food they picked and delivered to households. The partner stores may even have gained some business in these cases. But, picking groceries from a retail store, with its own markup already on the product only raised the cost of Internet selling, a cost that consumers were not willing to pay. The next step was for Internet sellers to establish their own distribution centers where cost-of-goods-sold is lower and groceries could be picked faster. But the capital investments in real estate and inventory and equipment also added to their costs. So, Internet food companies lost money continuously because even as they increased sales and revenue.

The fixed and variable costs of procuring and servicing every new customer was far greater than the revenue generated. One reason it is so hard to make a profit in this business is that Internet food companies lack the power of volume buying that is enjoyed by large food retailers. They simply cannot obtain the goods they sell at the lowest prices. Another is that consumers become dissatisfied with a merchandise mix that does not have enough variety. And finally, at a delivery charge of $10 per order, covers only about 60 percent of the delivery costs.

Financial analysis in April of 2000, by the Boston Consulting Group estimates the costs of obtaining each new customer to be $82 for an Internet-only retailer, $31 for a store-based Internet retailer and $12 for a traditional retailer during 1999. By the second quarter of 2000 the same consulting firm estimated that e-tailers spent only $40 to obtain each new customer. Table 1 shows an estimate of the percent of sales revenue devoted to covering costs and the gross and net profits of supermarkets, retail Internet companies that have their own distribution centers, and category killer Internet companies that buy directly from manufacturers.

Table 1. Possible Operating Statements for Retailers with various Formats

	Supermarkets	Internet Seller with Own Depot	Category Killer Internet
Seller			
Sales	100%	100%	100%
Cost of goods sold	70.6%	76.0%	64.0%
Gross Margin*	29.4%	24.9%	36.0%
Store	21.4% (includes 7-11% for labor)	NA	NA
Order entry and customer service	NA	NA	1.5%
Warehouse/ Depot	2.5%	3.1%	2.7%
Delivery	1.0%	10.9%	14.8%

Marketing and Advertising	1.5%	3.1%	12.5%
Operating profit	2.1%	2.6%	3.0%
Depreciation and Taxes	1.1%	1.0%	0.6%0
Net Profit**	1.0%	1.6%	1.6%

The category killer can buy merchandise directly from manufacturers but spends considerably more in marketing and advertising. The delivery costs for Internet companies approximately off set the labor costs for stores. This has led to the traditional bricks-and-mortar companies buying up Internet food retailers. David Ignatiou, a reporter for the Washington Post, calls it the "revenge of the dinosaurs". On April 14, 2000, Ahold, a Royal Dutch parent company of Giant Food, bought 51 percent of Peapod with the right to purchase up to 75 percent. April 17, Safeway announced it will buy 50 percent of Groceryworks.com. In a model similar to large pharmaceutical companies purchasing patents for new drugs that come out of basic research laboratories in universities, large traditional food retailers are buying up Internet companies with proven software and logistics systems that were financed by venture capitalists and developed by technology wizards.

The large bricks-and-mortar retailers can spread the costs over far higher volumes and, they have the brand recognition and consumers' trust that Internet retailers do not enjoy. It is widely believed that business to consumer Internet sales will be dominated by "bricks-and-clicks" companies in the future but finding the right mix of integration and separation will be a challenge to individual companies. Though none of the products listed was food, the ratio of off line dollars spent to online dollars ranges from 2.92 for clothing to .68 for books. This implies that much of the on-line shopping is just that - shopping (searching) for items and descriptions and prices that enable a consumer to go to an offline source and buy. If this is the best use of the Internet, then a bricks-and-clicks combination looks like a viable format.

In a demand driven system of food sales, long term success will depend on consumers' adoption of Internet shopping. In economic parlance, if it reduces their search costs and increases their utility (delivers superior quality products at lower time and money costs) it will be used. If it costs them time, hassle, choice, or variety they will return to the bricks-and-mortar store or they will obtain food from a food service place.

Economic theory of consumer behavior predicts that as household incomes rise and the value of time increases, consumers willingness-to-pay for the costs of food delivery will increase. But the value of the service and the quality of the food delivered via an Internet seller must exceed that which can be had by shopping for oneself if it to be a sustainable business. In a study of why Internet shoppers come back, the number one reason was level and quality of customer service followed by on-time delivery. Price was the last of eight other reasons .

Another study found that consumers top reasons for shopping on line were: a.) they can shop anytime, b.) it takes less time than going to stores, and c.) they dislike holiday crowds. Reasons they were disappointed with the online service were: a.) holiday merchandise arrived late, b.) they had to pay extra to ensure on-time delivery and c.) they received only part of an order . All of these reasons indicate the importance of saving consumers' time and delivering quality products and service to retain Internet shoppers.

The taxonomy developed for electronic retailing focuses on the digital content of the product-service offerings and the market segment it intends to serve. It sets a standard for how to examine this type business and it informs us about the state of the art in electronic food retailing. They found that even heavy users of dynamic digital content on their Internet sites do not deliver many value-added service-products to their customers. Unlike some of the business to business Internet exchanges, these retails do not aggregate diverse suppliers of food and make them available to consumers, at least not more so than a typical grocery store. If anything, the variety available is less than in a bricks and mortar store.

Business to Business E-Commerce

The second, and growing, use of e-commerce is for market discovery between buyers and sellers in the supply chain, typically before the final consumer. These "online market makers" are fundamentally different from the retailer/supplier relationships that will be discussed in the next section. There, an intimate business relationship develops using information technology to cooperatively manage the inventory flow through the store. That relationship is contractual and expected to last at least through the next selling period. On-line marketplaces facilitate shopping by buyers at all stages of the supply chain.

These market makers rarely own any of the merchandise that is traded through them, they simply help buyers find the best price or value available and help sellers identify buyers. Since there are a large number of diverse producers at one end of the food supply/demand chain and a large number of diverse retailers at the other end, this model fits the food industry very well. Farmers may shop for seeds and fertilizer on line. Farm input suppliers may have their own WEB page to sell online, or dot-com companies that specialise in selling farm inputs will arise and act like a supermarket for these type of products. In fact, many agricultural.com companies have sprung up.

Recognising them as competition, three large companies announced an alliance to create Rooster.com through which to market their products to farmers and others. The vulnerable businesses from this type of Internet selling are the local farm supply stores, seed and feed mill and smaller startup companies trying to sell supplies on line.

A company called Freemarkets, Inc. conducts auctions on line for businesses to find and purchase parts and ingredients and equipment. Numerous alliances of companies seeking to sell their products to other businesses or consumers are developing in order to take advantage of economies of size and skill in setting up such sites. Many are announced. Few are operational at this time. Consumers discovered a product on the Internet and its sale increased beyond all expectations of the seller. Market discovery is one of the fondest dreams of Internet companies as witnessed by their aggressive advertising budgets."

An example of a reverse aggregator with a reverse auction market for consumers is Priceline.com which brings together consumers and products in their retail grocery store at a lower than average price. Consumers bid on-line for the lowest price the exchange will accept for a given grocery item. Once it is accepted, the consumer pays Priceline with a credit card, receives a printout of the purchases that they take to their local retail food store (that has agreed to participate with Priceline.com) and picks up that item. The retailer gets paid by Priceline. This is clearly geared toward the low price shopper and like most other new e-hubs, its longevity will be tested by how well it adds value to the shopping experience of consumers and how efficiently it serves all the parties involved.

Business to business e-commerce is nothing less than a new "way of doing business." It tends to follow a "reverse product cycle" where process efficiency gains come first, followed by quality improvements to existing products and services, and then the creation of new services and products appear. It is a platform from which new institutions and products emerge. The food industry has traditionally created new (branded) food products within the research and development divisions of food manufacturers, tested their sales in selected markets, advertised them heavily, and offered them to wholesalers/retailers at deep discounts or with slotting fees. It is a system with high transaction costs and risks of rejection by consumers.

With e-commerce systems, consumers help create the products through their interaction with the sellers. Business to business e-commerce, as it is being adopted by retail food stores and their suppliers, has focused mostly on ways to save labor costs, speed up ordering, delivery and invoicing, and move the product through the system as fast as possible. The latest innovations have occurred because of new electronic technology, information management systems, and new competition. Even though food retailers were early leaders in the development and design of universal product codes (bar codes) they are among the last to realise the payoff from their universal adoption and use.

With compatible computer systems and the willingness to share data with suppliers, the information about what was moving over a scanner in a store could be transmitted directly to suppliers - their own distribution centers and/or manufacturers. Manufacturers could, in turn, adjust their supplies (or production lines) according to consumer demand aggregated from each store. Theoretically, by making information about sales at all retail stores available to both the retailer and their suppliers simultaneously, a continuous loop was created whereby information about sales flowed in one direction and products flowed back, just in time to match the retail demand.

The concept of sharing information about sales with vendors and developing a continuous and coordinated flow of products was introduced to the rest of the retail food industry and institutionalised by a coalition of trade associations and food manufacturers and suppliers and a few big chains under the name of Efficient Consumer Response (ECR) in 1992. It had little to do with the consumer except that its goal was to track consumer purchases at the point-of-sale and share that data with suppliers so they could tailor the delivery of goods to match the volume being sold. The goal of ECR was to have each food store/company behave like Wal-Mart; to implement electronic data interchange (EDI) to order goods and slim down the offerings in each category in order to streamline delivery and the costs associated therewith. This leads to "category management" which has had considerable success even though it may conflict with a goal of providing variety and service to consumers.

In 1998, 24 percent of stores responding to a survey by the Food Marketing Institute reported using EDI with at least some suppliers. Of those who did, 53 percent used a third-party, value-added network (VAN). This is a network that connects different members of a retailer's supply chain using Web-type technologies and interfaces. Only 17 percent were using the Internet and the rest used both. A major stumbling block to adopting management practices advocated under the umbrella of ECR is that EDI requires compatible computer systems which are expensive to set up and operate. ECR suffered from a lack of what economist's call "network effects".

Network effects occur in markets for systems when the value of a system to one user is positively affected by another user joining and enlarging the system. Market behavior and performance of electronic services of many types are profoundly and positively affected by the adoption by a critical mass of users. Two examples are the usefulness of fax machines and e-mail. Adopting new network technologies also depends on the expectations about whom else is in the network and how much joining it will lower costs and increase profits. The more other users one expects to find in the network the more likely they are to join. As the number of users of a network grows, the benefits to each user grow above the price they paid for joining the network.

In economists terms, the social benefits rise above the price paid by users and there exists a classic case of positive externalities and the network begins to look like a public good; its use is non rival and nonexclusive. If in addition, as the number of users rises, the average and marginal costs of providing network services fall, the network could be in the position of being a natural monopoly in the same way we formerly thought of electric power utilities and telephone companies as being natural monopolies. The development of e-commerce is a long way from this stage in 2000, but the vision of vertically integrated supply chains points in this direction. The network, like the old fashioned phone lines, would provide the compatible, ever ready, and seamless communications between retailers and manufacturers or other suppliers.

The technical problems of incomparability and a cultural resistence to sharing store level data with suppliers resulted is very slow adoption. As expected, the largest chains adopted first. Large chains all had their own distribution centers so they did not have to communicate with a wholesaler standing between them and a manufacturer/ vendor. Self-distributing chains were only slightly ahead of the multi store, non-self-distributing chains on the building of relationships with suppliers.

These findings support the positive "network effects"; using information technology both allows and demands larger sized organisations and networks to be effective at lowering costs and increasing logistic efficiencies.

Smaller retail stores were simply not willing or able to participate in electronic data interchange (EDI) necessary to participate in an efficient response relationship with suppliers. But, apparently the largest chains, already in supplier networks, believed that there was industry wide economics of scale to be gained if more retailers and suppliers could be convinced to join. In other words, they envisioned the benefits to the whole industry and themselves if more retailers and manufacturers could be enticed into the "network."

With the historical record of consumer sales, the manufacturer and the retailer each forecast sales over some future time period, share their forecasts, and negotiate anticipated future sales if necessary. Manufacturers agree to deliver merchandise on a prearranged schedule and manage the inventory of their products in each store. This system obviously demands accurate scanning information and some Internet interface over which data can travel securely. It also demands a willing-ness to share data and the responsibility for the products on the shelves.

Businesses to business e-commerce network systems involve noth-ing short of re-engineering the business process, changing the culture, and integrating data from one place to another; from a retailer's sales floor to a decision system that involves a manufacturer, somebody's warehouse and a transportation system, and being able to evaluate and change options on the fly. He further says that if the markets become as efficient as planned, it will be a very hard way of life. There will not be as many profits to go around and managers will have to find new ways to make money in a truly competitive world. This is consistent with the theory of network creation and network effects. As everyones' costs decline in a large efficient network, competition will increase and new networks will arise to define unique niche markets.

As e-commerce models and capacity made economies of scale and scope more obvious the largest retailers began to merge in unprecedented numbers and size. At the beginning of 1998, the United States had no nationwide grocery chain. In August, 1998, Albertsons acquired American Stores becoming the first. The west coast had already been consolidated under Fred Meyer, Inc. and in 1999 Kroger bought Fred Meyer to become the largest nationwide grocery chain. The rational for these mergers was the increased buying power larger chains would have with manufacturers. This could drive down the cost-of-goods-sold in order to compete with Wal-Mart. The national concentration ratio of the top four retail food chains had been remarkably stable at about 16-17 percent of sales for decades; in 1999 it rose to 34 percent. The top Five make up over 40 percent of retail food sales.

The concentration ratio at retail is rising and in individual cities it is not unusual to have ratios of 70-85 percent. With bricks-and-mortar retailers buying up Internet companies, that can only increase. In the end, whether this will lead to more efficient supply chain management and lower costs of food for retailers and consumers remains to be seen. Many studies have shown that when concentration grows, the average cost of food grows also. Translating efficient supply chains into price reductions for consumers requires sufficient competition. Factors that can inhibit competition in an e-commerce era are sector-specific transaction structures, the advantages gained by the first-movers, and differences in the regulatory environment between regions or cities. Driving costs down will not be enough for the new e-commerce era to improve the well being of firms and their consumers.

Widespread disintermediation is occurring as business to business e-commerce allows consumers and/or retailers to communicate directly with manufacturers . The most vulnerable types of firms are the wholesalers and brokers. They too have been consolidating, disappearing or changing their business portfolios. Traditional, third-party wholesalers share of the intermediary business has dropped from around 42 percent in 1990 to 38 percent today. Wholesale firms, such as Flemming, Nash Finch and Supervalu, provide wholesale services to thousands of smaller, independent stores scattered all over the country. In order to survive an increase in manufacturers' direct store delivery (DSD) activity (now 28 percent of the intermediary sales) and the growth in the portion of the market taken by ever growing self-distributing chains' distribution centers (now 34 percent of intermediary sales), they have merged and begun to acquire more retail stores as part of their business.

Some of the largest, such as Supervalu consider themselves to be "virtual chains." Some stores are company owned and some are franchised, and some are simply supplied and serviced in a traditional fashion. These large wholesale suppliers have established their own e-commerce networks for ordering and category management with limited success. They face the same constraints that prevented electronic data interchange between retailers and their suppliers in the earlier days of ECR. Small independent stores have neither the capital nor the incentive to invest in electronic data systems and easy access networks are not yet available to them.

The economics of self-distributing retail chains' distribution center is convincing. When there is an agreement with a manufacturer to replenish shelves according to the sales demand, many manufacturers choose to deliver it themselves, stock the shelves, manage the inventory and, in some cases, retain ownership of the product until it is sold, that is, scanned upon purchase. In this case, the retailer has no cash tied up in inventory, they own fewer assets, they save on labor, and they have more cash to use between the time the item is sold and they must pay in manufacturers' invoice. Salty snacks and beverages and some bakery products are most likely to be delivered directly and handled on a "scan based trading" basis. It could be argued that this is one way for manufacturers to regain control over their products and retrieve some of the bargaining power that flowed to retailers as they held the key to system wide inventory management, namely the scanner data.

Future Challenges

E-commerce is a relatively new phenomenon. Use of the Internet has penetrated over 25 percent of U.S. households in less about 5 years, the personal computer in about 15 years. Ways to study the rapidly evolving products and services that emanate from the use of the Internet are not immediately obvious. Just as the Internet brings us a reverse product cycle, where the process comes before the product, the business models, startups, failures and successes have to occur before most academics can know how to think about them. Recall that it is a new way to do business. That means that we need new ways to apply theories and new models to study the new market behavior.

The development and success of individual firms can be modeled as the optimum mix of products and processes. Household economics and the vale of time would be a useful framework to study consumers' likelihood of adopting Internet shopping and the mode of delivery they choose. It leads to broader questions of tradeoffs between labor and capital in other segments of the supply chain considering adoption of e-commerce options.

Other economic theories that might be used to explain and predict firms' behavior in the supply chain were proposed by researchers. They focused on vertical integration that would ensure consistent and predictable quality in the inputs to food manufacturers, but one can envision how each of their theories might apply to e-commerce behavior closer to the retail end of food markets. For example, transaction cost economics examines and predicts how organisations choose governance structures in order to minimise transaction costs.

It can be used to explain greater vertical coordination where capital investments are expensive (developing hardware and software to collect and use very large sets of data), highly specific to the industry (food sales in grocery stores where all goods do not have bar codes), and the level of uncertainty about the quality of a product is high (seasonality of fresh produce, inability to judge quality by inspection, and a huge number of small, diverse, and uncoordinated sellers). A shifting of power within a supply chain as e-commerce develops is a curiosity and a concern. Transaction cost economics and other models dealing with principal-agent behavior might be used to explain the sharing of risk between retailers and manufacturers who share data and responsibilities for who owns merchandise in the store, who decides on displays, inventory depth, and category management.

Principal-agent models examine how the principal party (the one with the power such as an employer) incents the agents the buy from or sell to, to behave in a way that maximises the returns to the principal. They typically treat technology as fixed, but given the institutions of e-commerce, how various parties emerge as principals and how they share the risks and incentives with their agents could be fertile ground for research. Ownership patterns and who has the right to benefit from ownership is the purview of property rights theory. It might be used to assess whether or not integration of two or more segments under common ownership will improve system-wide performance . It comes into play with private label products, with the sharing of scanner data, and with the use of consumers private demographic and buying pattern information.

Changes in the market structure imply that businesses will be larger and be involved in more strategic alliances to control and/or coordinate the supply chain for production inputs and final food products. Alliances of major suppliers of raw materials will sell to manufacturers over Internet platforms that allow information to be shared with its members, or at least, between buyers and a group of sellers. These large alliances are being formed to short circuit several small startup Internet brokers all the way from farmers, to manufacturers, to the retail distribution centers (warehouses).

The human and physical (electronic) capital necessary to participate in one of these large alliances encourages and demands that firms grow larger. They favor the larger retailers and encourage further consolidation into self-distributing chains. Large self-distributing chains are in a better position to form partnerships with large manufacturers to share information about consumer sales, cutting transaction and distribution costs in the system. These networks will probably create barriers to entry for non mainline grocery businesses, but in this market there is plenty of room for niche, neighborhood, and regional players. Food is needed in every local. People like to select local, fresh, and natural foods. They are willing to pay more for variety, experiences and service with their food and large, low cost operators do not excel at providing these characteristics.

The personal touch, the local flare, will be provided by local food retailers and the food service sector. However, to the extent that there is profit to be made in selling local, fresh and natural foods, large retailers have an incentive to form buying alliances or contracts with producers who will guarantee these quality characteristics. The value added to food through being able to trace a particular food product to its production source includes developing trust by consumers who value safe, high quality food. Therefore, the industry structure has reacted with increased horizontal consolidation and increased vertical cooperation. The network effects of cooperation are powerful incentives to share price, quantity(sales) and quality information between buyers and sellers in the supply chain. Realising the incentives and the payoffs is slow in coming since it involves nothing less that an cultural change among thousands of managers and workers.

E-commerce alters the concept and use of time. There are faster production and delivery cycles and consumers can shop at all hours of a day. Understanding the changing concept of time and how activities of work, household tasks, and leisure blend together will be another challenge for meeting consumer tastes and preferences in this new market. Being continuously connected to your work, your friends and family, and the marketplace provides unlimited opportunities for interaction but little time for reflection and synthesis. New status symbols involving freedom from connection will surely arise.

Just-in-time delivery models, born of business to business e-commerce and information sharing, have their own time constraints. The physical infrastructure, such as highways that are congested, and unpredictable demands due to weather emergencies, make lean, efficient delivery systems vulnerable to default. Increases in productivity in the entire economy are evident since the advent of Internet and computer technology. Going from a predictable and desirable two percent increase in productivity to a run of years with over three and four percent speaks loudly for the new ways of doing business.

Exactly where these increases are coming from, which firms or which sectors, is less clear. There is some evidence that costs are lower and sales higher in retail and wholesale food firms that have adopted advanced electronic and information technologies and who have built strategic alliances with their suppliers and customers. This is not to say they are necessarily more productive or that they bring greater satisfaction to their consumers. Measuring productivity and well as general social welfare increases is a challenge with the advent of the Internet.

5

Microwave Food Heating

The term "microwave" generally means a microwave oven, which is used in many households for heating food. The microwave oven has been an astonishing success, now about 95% of homes in America have at least one microwave oven. Microwave ovens are generally used as convenient method for reheating rather than cooking.

In microwave heating, the heat is generated due to the molecular friction between dipole molecules (example: water). Since water is easy to stimulate and most of the food products contains 40-80% water, microwave heating is highly suitable for cooking. Salts, fats and proteins also act as dipolar components and affect heating rates. Thus, microwave penetrates food easily and heats the food from inside out. Drawbacks of microwave heating are its inability to brown food, non-uniform cooking, and excessive drying of foods such as breads. Another major problem in microwave cooking or reheating of the food is the chance for pathogen survival.

Population groups such as pregnant women, immunocompromised, elderly, and young children are highly susceptible for food borne infections. As the cooking or reheating of the food is one of the last steps in food preparation, it should assure food safety. Therefore, food safety becomes important and cooking should ensure food safety. Though the probability of pathogen survival in microwave-cooked food is higher than that for conventionally cooked food, consumers increasingly relay on microwave cooking.

The reduction in process time, often as much as 10 to 1 compared to conventional methods, and higher food quality in terms of its appearance, flavor and 2 taste, convenience, easy clean-up, decrease in cost of electricity, more retention of vitamins and minerals can be main reasons for consumer preference to microwave cooking. Therefore, elimination of pathogens during microwave cooking becomes very critical. Understanding the microwave cooking process, and the changes in the properties of food during cooking are important issues in developing specific recipes and heating instructions to assure food safety.

Mathematical models, simulating the microwave-cooking process, play an important role in designing and optimizing the cooking process. Thc modeling technique provides an insight into the microwave cooking process. Product developers will be able to test the effects of the changes in the food formulations without delay or with out the cost of having to work on product samples.

Thus, the mathematical models are useful in improving the microwave cooking procedure. Using modeling technique, temperature profile in a microwave-heated food can be predicted. This could help in understanding the cooking process and studying various factors that affect the microwave cooking. This can lead to heating procedure, which will assure food safety.

Microwave Heating of Food

An increase in temperature in a food or any material can

be achieved in three possible ways, namely, conduction, convection and radiation. In conventional cooking methods, food is normally heated by conduction and convection where as heating through radiation uses radio frequency waves, microwave, and infrared radiation. While the radio frequency heating is still under development, infrared radiation is employed for surface heating, as the IR waves cannot penetrate below the surface. Microwave, on the other hand, penetrate food and produces heat from inside out.

Microwave oven is one of the most commonly used appliances for heating food. Almost every home in North America has at least one microwave oven. Most of the problems associated with microwave heating occur due to its uneven heating nature. Though there have been problems, consumers still use microwave oven for its fast and easy use. It is highly suitable for busy life style as it can heat food much faster than other conventional heating methods. Understanding how the microwave heats the food and the science behind the microwave would be helpful in reducing the problems associated with microwave cooking.

Microwave is generated by magnetron. Magnetron is a circular symmetric tube like diode that consists of a cathode as the central axis of the tube and an anode around the circumference. The magnetron contains a space called resonant cavities. Resonant cavities act as tuned circuits and generate electric fields. These cavities also determine the output frequency of the microwave. The magnetron has an antenna connected to anode and it extends into resonant cavities.

Antenna is used for transmitting the microwave from the magnetron to waveguide (Figure 1). The magnetic field is created by a magnet that surrounds the magnetron. When power is supplied, an electron-emitting material at the cathode becomes excited and emits electrons into the vacuum space between cathode and anode. The energy of the electrons is caught in the fields. The excess microwave energy travels as waves and extracted by the antenna. The antenna transmits the oscillating waves to the waveguide where they travel into oven cavity. The waveguide is a hollow metal tube.

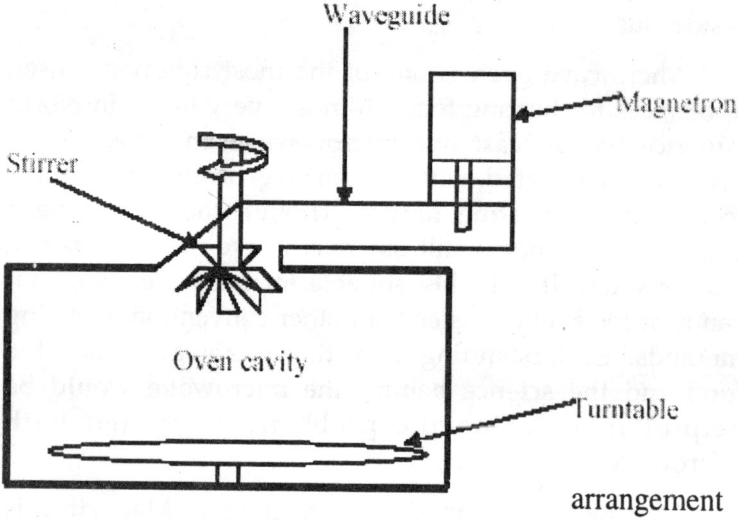

Figure 1. Major components of typical microwave oven

Metallic walls of the waveguide are nearly perfect electric conductors and microwave propagates with low transmission losses. As the waves enter the cavity, they are dispersed by a stirrer (Figure 1). This action minimizes hot and cold spot in the oven cavity. Normally, the magnetron operates with efficiency around 60-65%. Once inside the microwave oven, waves can be reflected off the oven sides and floor, can be transmitted through containers and lids made of glass, paper, plastic, and can be absorbed by medium such as food.

The electromagnetic waves, which includes microwave, travel at the speed of light (3.108 m/s) and possess energy in the form of high-energy packets known as quantum energy. The quantum energy can break the chemical bond when the quantum energy exceeds the chemical energy. Gamma rays and X-rays, which have short wave lengths, high frequency and high energy, are capable of breaking the chemical bonds. Microwave and radio waves, which come under long wavelength, low frequency, low energy category, do not have enough energy to break chemical bonds.

Microwave belongs to the group of non-ionizing form of radiations, since it does not have sufficient energy required for the ionization process. The quantum energy in the microwave is responsible for creation of heat as the microwave oscillates 2450. 106 times per second and the dipole molecules align to the electric field of the microwave at the same rate. The alternating electric field stimulates the oscillation of the dipoles of the molecules (example: water) in the food. The heat is generated due to the molecular friction between dipole molecules. Breakage in the chemical bond might occur due to this heat generation, not due to microwave directly.

Electric field component of microwave is responsible for heating. It causes the molecules of dielectric materials to rotate, and produces rise in temperature due to friction between molecules. Magnetic field component of microwave does not take part in heating food. However, some susceptor materials in the food package may interact with magnetic field. Inside a microwave oven, when incident and reflected microwaves interact, standing wave patterns are formed. The standing waves patterns have maximum and minimum (when incident and reflected waves cancel each other) value at certain distances from the reflecting surface. It should also be noted that incident and reflected waves are always in continuous motion.

Inside a microwave oven, there are high possibilities for multiple reflections and hence there are number of standing wave patterns. Every possible pattern is referred as a mode and at least 20-30 different modes are possible in a typical microwave oven. Therefore, the microwave oven cavity can be termed as multimode cavity. When a load (food) is placed inside a microwave oven, it changes the reflections and hence the standing wave patterns and the modes.

Sources of Heat Generation

Microwaves inactivate microbes mainly by thermal effects. The quantum energy level in microwave is very low and the chances for its lethal effect on microbes and production of toxic compounds in the food are very remote. The microbial inactivation by microwave depends on the same time-temperature relationship as seen in conventional heating methods. The microwave heating occurs instantly through out the product compared to conductive heat transfer from surface to interior in a conventional oven. Therefore, microwave heating is much faster and the temperature necessary to kill microorganisms is reached quickly.

The rotation of dipolar molecules accounts for the most of the heat generated during microwave processing. These dipolar molecules try to align with microwave field at a speed consistent with the microwave frequency. This rapid movement of polar molecules results in the development of heat, because of the friction between molecules. Another mechanism responsible for heat generation is ionic polarization. Ionic polarization occurs when ions in solution move in response to the applied electric field component of microwave. Ions are accelerated by this electric field. Displacement of ions causes collision with other ions, converting kinetic energy into heat. At microwave frequencies, numerous collisions occur, and much heat is generated. However, it is a less important mechanism than dipole rotation.

Characteristics in Microwave Heating

The important food characteristics are dielectric properties of the food, thermal conductivity of the food, size, shape, orientation of food in relation to oven, physical state of water in the product, the presence of bone in the food and its moisture content.

Dielectrics properties

Microwaves interact with materials based on their electrical properties. Metals are good electrical conductors and good reflectors of microwave. They are not heated by microwave. Materials such as dielectrics come under electrical insulators and are good absorber and transmitter of microwave. Heat is generated in the dielectric (or food) primarily through absorption of microwave and the absorption, and hence the dielectric properties, depends on microwave frequency, food composition, product temperature, physical state of water in the food, and product density. Absorption characteristics of a food can be changed significantly by altering the above factors, for example, addition of salt in the food increases absorption of microwave.

Microwave heating is mostly dielectric in nature and involves the rotation of dipolar molecules. The dielectric properties of a food are characteristics of the materials determining the interaction of electromagnetic energy with the materials. The dielectric properties such as dielectric constant and the dielectric loss factor play a major role in microwave heating. The capability of food to store electric field energy is dielectric constant. The dielectric loss factor measures the ability of food material to dissipate electrical energy as heat.

Therefore, the amount of dipolar molecules significantly affects the heating. When the dipolar food components are not evenly distributed in the food, uneven heating can be expected. Obviously, the difference in dielectric activity is a common problem in foods with more than one ingredient. Dielectric properties for some of the common food materials are presented in figure 2.

The measurement of dielectric properties in microwave frequencies can be done using waveguide and coaxial transmission line method, Open-ended Probe Method, and Cavity Perturbation Method. Generally, measurement techniques can be categorized into reflection or transmission measurements by resonant or non-resonant systems, with open or closed structures for measuring the dielectric property of the material.

Closed structures method can be divided into waveguide and coaxial line transmission measurements and short-circuited waveguide or coaxial line reflection measurements. Open structure techniques include free space transmission measurements and open-ended coaxial line measurements. Resonant cavity structures can be closed resonant cavity or open resonant structures. In the case of open resonant structures, the measurements can be done as two-port device for transmission measurement or as one port device for reflection measurements.

Thermal conductivity of food

The thermal conductivity of food plays important role in microwave heating. High thermal conductivity materials dissipate heat faster than low thermal conductivity materials during microwave heating. Food with high thermal conductivity will take lesser time to attain uniform temperature during holding period. Specific heat of food determines how fast a food can be heated. Specific heat can be raised by increasing solid content by adding

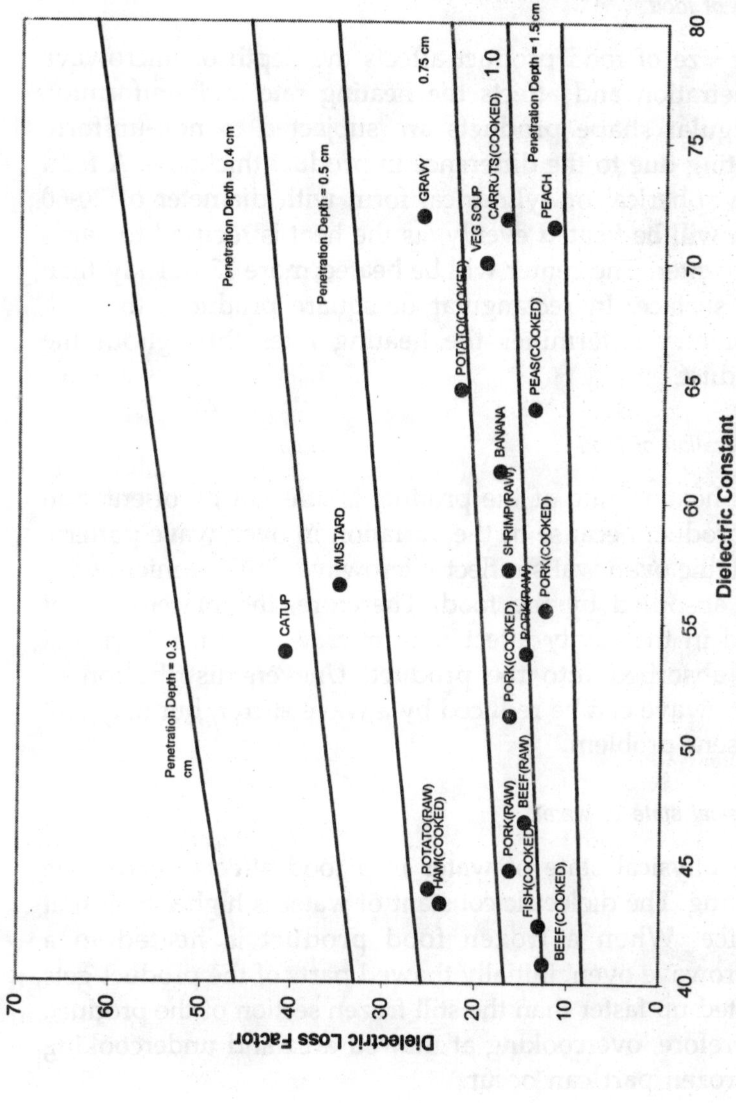

Figure 2. Food map for dielectric properties

components like salt, protein. Specific heat along with thermal conductivity constitutes thermal properties of the material.

Size of food

The size of food product affects the depth of microwave penetration and affects the heating rate and uniformity. Irregular shape products are subjected to non-uniform heating due to the difference in product thickness. A food of a spherical or cylindrical form with diameter of 20-60 mm will be heated evenly, as the heat is focused towards the center. The center will be heated more 13 quickly than the surface. In rectangular or square products, the slab geometry determines the heating rates throughout the product.

Orientation of food

The heating rate of the product is affected by orientation of product because of the variation in oven wave pattern. Metallic oven walls reflect microwave until all microwave are absorbed by the food. Therefore, the placements of food in the cavity affect how microwave hit, reflect and are absorbed into the product. Uneven distribution of microwave can be reduced by a wave stirrer, but may still present problem.

Physical state of water

The physical state of water in a food affects microwave heating. The dielectric constant of water is higher than that of ice. When a frozen food product is heated in a microwave oven, initially thawed parts of the product gets heated up faster than the still frozen section of the product. Therefore, overcooking at thawed area and undercooking at frozen part can occur.

Moisture content

The moisture content of the product significantly affects the dielectric properties of food and consequently, the penetration depth of microwave is affected. Uneven heating rate is observed in high moisture foods because of low microwave penetration depth. Low moisture foods have more uniform heating rate because of deeper microwave penetration. In addition, in high moisture foods, heat loss 14 through surface cooling can occur as a result of evaporation.

Microbial safety

Food safety of microwave cooked is critical in developing cooking instructions. The cooking instructions on the package must ensure that the pathogens are completely eliminated. Inoculation studies by researchers showed that the food could sustain pathogen survival when the temperature is not closely monitored in the food product. A study on ground beef loaves concluded that fat content of food plays an important role in the pathogen survival. Higher fat content of ground beef helps in reduction of pathogen survival.

Microwave cooking of vegetable product has higher risk of pathogen survival than that of meat products. Post process contamination is contamination that occurs after cooking food due to contact with unsanitary or uncooked materials. Improper handling of food increases the chance of pathogen survival. Allowing holding time after microwaving could decrease the microbial population. However, many consumers do not realize that proper holding time can considerably increase food safety.

Storage abuse is another factor to be addressed on microbial safety. Abuse in storage temperature whether it is in food industry, grocery store or consumer home would increase the pathogen survival. High storage temperature encourages growth of microorganisms. In addition, damage to the food or food packages during shipping and handling could be serious threat to food safety.

Reducing Non-Uniform Heating

Microwave heat food unevenly due to uneven distribution of power. Hot and cold spots exist in the food due to uneven wave distribution. Developing instructions for cooking food is very challenging because of this uneven heating. Understanding the non-uniform heating with respect to oven, food, container shape and geometry, and the dielectric properties is very critical and mathematical modeling is a vital tool to do this.

More uniform cooking results when food is positioned above the floor (for example, on a glass plate) than when food is placed on the floor. When the food is positioned above the floor, the microwave can then be reflected off the wall and floor of the oven, transmitted through the plate and be absorbed on the bottom of the food. To increase the effect of uniform heating, microwave ovens are normally equipped with a mode stirrer. The mode stirrer is just a metallic fan blade, which is used to perturb the field distribution inside the oven. The food is also rotated with a motorized platter to increase the uniformity of heating by reducing concentration of power at certain places in the food.

Providing sufficient holding time after the microwave heating gives time for heat conduction in the food and more uniform temperatures results. Post-processing temperature rise (PPTR) is the increase in temperature after microwave heating as measured at the coolest point in the product. PPTR might depend on both the product and power level and help to reduce the differences in temperature with in the product.

The provision of power settings in some of 16 microwave ovens also helps in more uniform heating of food. When a microwave oven is operated at a power level other than full power, the food undergoes heat-hold-heat-hold cycle. The holding time in this case is determined by the power level chosen. Therefore, reduction in temperature differences in the food can be expected.

Modeling of Heating Process

Modeling represents a phenomena using set of mathematical equations. The solutions to these equations are supposed to simulate the natural behavior of the material. Modeling can be a design tool to develop food that will provide optimum heating results in the microwave oven. In the modeling work, the food system is represented as being made up of many small elements in the simulation process. These discrete elements are joined together to make up the product.

Modeling of microwave cooking process can involve two separate parts, one being modeling of heat and mass transfer in the food and another being modeling the electromagnetic field inside the microwave oven cavity for calculating heat generation term.

Modeling of electromagnetic field arises when Maxwell's equations are used for calculating the heat generation term. Maxwell's equations are the basic laws for the microwave propagation. If the model employs Lambert's law for calculating the heat generation term, electromagnetic field is not modeled and the simulation work becomes easier.

Fat Content and Temperature on Dielectric Properties of Ground Beef

The dielectric properties describe the behavior of the material when subjected to electromagnetic fields for dielectric heating applications. These properties are important in determining the penetration depth of microwave power and power absorption rate by the food in a microwave oven.

Dielectric properties of interest are dielectric constant (ε") and dielectric loss factor (ε"). Dielectric constant is a measure of ability to store electrical energy. Dielectric loss factor is a measure of ability to convert electrical energy to heat. Dielectric constant and dielectric loss factor can also be defined in terms of relative complex permittivity, $\varepsilon = \varepsilon'' - j\varepsilon'' = |e|\, e^{-jd}$, where $\tan \delta = \varepsilon''/\varepsilon'$ is termed as the loss tangent or dissipation factor. The magnitude of these parameters determines the interaction of the microwave field with the material.

Dielectric properties of food and agricultural materials are important in developing microwave heating procedures and in the design of electrical and electronics equipments that interact with the food. They can be related to the composition, which is useful in quality control and assurance. Dielectric properties can also be used to analyze the biophysical properties of the composition as the temperature changes. They will be helpful in determining proper microwave heating time in automatic smart kitchens.

Consumption of ground beef in the US was 3.2 billion Kg in 1995. Per capita consumption of beef increased from 29 Kg to 30 Kg between 1990 and 1998. It is important to have well - developed microwave heating process for cooking ground beef. Measurement of dielectric properties at various temperatures and at various fat levels will help to achieve well-developed microwave-heating procedures and also help in understanding behavior of the ground beef during microwave heating.

Figure 3 shows a schematic diagram of experimental apparatus. The dielectric properties determinations were

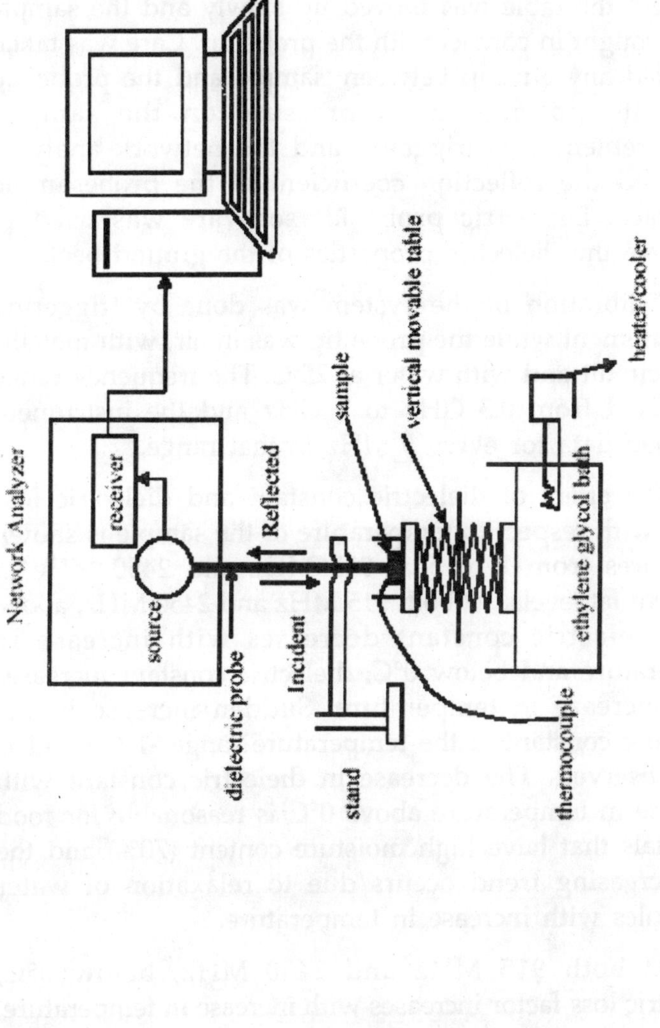

Figure 3. Schematic diagram of experimental apparatus

made by measurements on an open-ended, 3.6 mm diameter, semi-rigid coaxial line with copper conductors, connected to a network analyzer. The coaxial probe was fixed in a place well by a stand arrangement. The sample holder was placed on a vertical movable table.

The heater/cooler unit was switched on and desired temperature was set. After the designated temperature was reached, the table was moved up slowly and the sample was brought in contact with the probe tip. Care was taken to avoid any air gap between sample and the probe tip without applying much pressure on the sample. Measurement was triggered and the network analyzer recorded the reflection coefficient at the probesample interface. Dielectric probe kit software was used to calculate the dielectric properties of the ground beef.

Calibration of the system was done by triggering measurement while the probe tip was in air, with metallic short-circuit and with water at 25°C. The frequency range was fixed from 0.3 GHz to 3 GHz and the instrument obtained data for every 5 MHz in that range.

The effect of dielectric constant and dielectric loss factor with respect to temperature of the sample is shown in figures from 2 to 5 at 915 MHz and 2450 MHz at different fat levels. At both 915 MHz and 2450 MHz, above 0°C, dielectric constant decreases with increase in temperature and below 0°C, dielectric constant increases with increase in temperature. Sudden increase in the dielectric constant at the temperature range -1°C to +1°C was observed. The decrease in dielectric constant with increase in temperature above 0°C is reasonable for food materials that have high moisture content (70%) and the 39 decreasing trend occurs due to relaxation of water molecules with increase in temperature.

At both 915 MHz and 2450 MHz, below 0°C, dielectric loss factor increases with increase in temperature.

At 915 MHz, above 0°C, dielectric loss factor increases with increase in temperature. This can be attributed to increase in ionic conduction in the product as temperature increases. This also indicates that the increase in energy absorption could occur with increase in temperature at 915 MHz. At 2450 MHz, above 0°C, dielectric loss factor decreases with increase in temperature till 45°C and then starts increasing with increase in temperature till 75°C.

Increase in dielectric loss factor was observed with increase in temperature at 915 MHz. Increase in dielectric properties with increase in temperature at both frequencies, below 0°C, may be due to the increase in fraction of water as temperature increases. Dielectric loss factor at 915 MHz is less than that in 2450 MHz at any particular temperature.

At any particular temperature, dielectric properties at 915 MHz are greater than dielectric properties at 2450 MHz. This could be due to the decrease in wavelength of 2450 MHz waves compared to 915 MHz waves. The wavelength and hence the microwave frequency, affects the dielectric properties. Small variation of dielectric properties among replications was observed. This might be due to variation in the product density and difference in the applied pressure by the probe on the food surface.

The open ended coaxial probe was at room temperature and this could have influenced the measurements at temperatures below freezing point. Differences in temperatures of food and probe could be the contributor for variations in observed dielectric properties especially at frozen state of the ground beef patties.

Effect of Fat Content

The effect of dielectric constant and dielectric loss factor with respect to fat level of the sample is figures from 8 to

11 for 915 MHz and 2450 MHz. Decrease in dielectric constant and dielectric loss factor was observed with increase in fat content at temperatures above 0°C. In addition, the values of loss factor were in close agreement. Since the low fat ground beef has high moisture content comparing to high fat ground beef, the decreasing trend is reasonable. It also implies that moisture content of the product is more important than fat content.

At all temperatures below 0°C except -1°C, the dielectric constant and dielectric loss factor remains almost constant irrespective of fat content. This might be due to the frozen state of water. At -1°C, sharp increase in dielectric constant and loss factor was observed from 4% fat to 9% fat. This could be due to increase in ionic activities at temperature close to melting point of ice. The increasing trend in properties after 9% fat to 20% fat was lesser than that between 4% and 9% fat.

Penetration Depth

As expected, the penetration depth at 915 MHz was greater than that at 2450 MHz. Higher penetration depth of 915 MHz was reported by researchers. Above 0°C, at both frequencies, high fat ground beef had higher penetration depth than low fat ground beef. At 915 MHz, penetration depth increases till 10°C and then starts decreasing. At 2450 MHz, penetration depth increases till 20°C and then decreases slightly with increase in temperature. Below 0°C, penetration depth increased with decrease in temperature, that is, microwave power absorption decreased with decrease in temperature at both 915 MHz and 2450 MHz frequencies.

Regression Analysis

Stepwise regression procedure was used to develop predictive equations for dielectric properties at 915 MHz and 2450 MHz by Statistical Analysis Software (SAS). R^2 value between predicted and observed values was greater than 0.82 in all cases. Predictive equations were of the form:

$$\varepsilon' = a_0 + a_1 T + a_2 T^2 + a_3 (F \times T) + a_4 F + a_5 F^2 + a_6 (1/T) \quad (1)$$

$$\varepsilon'' = a_0 + a_1 T + a_2 T^2 + a_3 (F \times T) + a_4 F + a_5 F^2 + a_6 (1/T) \quad (2)$$

where, T is temperature in °C and F is fat content in %. Dielectric constant and dielectric loss factor can be determined for varying fat levels at different temperature using these predictive equations.

Effect of Plant Parts and Temperature

Like light, microwaves are composed of electric and magnetic fields (electromagnetic radiation). The key difference is the frequency. Microwave frequency generally implies electromagnetic waves of frequency between 0.3 GHz and 3 GHz Dielectric properties of food and agricultural materials are important in developing microwave heating procedure and in the design of electrical and electronics equipment that interact with the food.

The dielectric properties describe the behavior of material when subjected to electromagnetic fields for dielectric heating applications. These properties are important in predicting power absorption rate by the food in a microwave oven. Dielectric properties determine the penetration depth of microwave power and local power absorption rates. This can be useful in developing microwave oven that have pre-programmed cooking time for type of foods.

Rapid method of moisture content determination is also possible for agricultural products. The dielectric properties of water (ε' ~ 75 below 5 GHz) are much different from that of most dry foods (ε' ~ 5). Therefore, measuring dielectric properties of moist product can provide an indication of moisture content. Dielectric properties of interest are dielectric constant (ε') and dielectric loss factor (ε").

Dielectric constant is a measure of ability to store electrical energy. Dielectric 59 loss factor is a measure of ability to convert electrical energy to heat. The magnitude of these parameters determines the interaction of the microwave field with the material. Broccoli is a commonly used vegetable in the United States. Study by ERS, USDA reveals that American consumers consumed 0.9 billion kilogram of broccoli during 1998.

Per capita consumption increased from 2.7 Kg to 3.65 Kg. Heightened awareness among consumers of the association of broccoli with good health expected to cause high demand of broccoli in future. Frozen broccoli market follows slow but study increase in consumption trend.

Cooking by microwave oven is very common nowadays. Developing proper microwave heating procedure for broccoli products becomes necessary to assure food safety. Study of dielectric properties will help in fulfilling this goal and in understanding heating behavior of vegetable products under microwaves. Also, the properties will be useful in developing smart kitchen and microwave oven that had been programmed for heating broccoli for right amount of time.

Measurement of dielectric properties is generally done by cavity perturbation technique, open-ended probe, waveguide and coaxial transmission line method. Dielectric measurement of a material-under-test follows four simple steps:

1. generate a microwave signal at frequency of interest,
2. direct the signal at/through the material under test,
3. detect/measure the changes in the signal caused by the material,
4. compute the e of the material from these changes.

Step (1) to (3) can be performed with help of Network Analyzers. Step (4) can be preformed by a computer program to compute e.

Waveguide and coaxial transmission line method

The values of dielectric constant and dielectric loss factor are derived from transmission line theory, which indicates that these parameters could be determined by measuring the phase and amplitude of a reflected microwave signal from a sample of material placed against the end of a short-circuited transmission line, such as a waveguide or a coaxial line.

In order to use waveguide technique, rectangular samples that fit into the dimensions of the waveguide at the frequency being measured are required. An annular sample is needed for coaxial line method. The thickness of the sample should be approximately one-quarter wavelength within the sample. Therefore, preparation of optimal sample requires guessing the dielectric constant of the material so that the wavelength can be determined.

Waveguide technique has very narrow operating ranges and generally used for single frequency measurement. Coaxial transmission line measurement technique is somewhat cumbersome because the sample must be annular geometry. These techniques are more suitable to porous materials.

Open-ended probe method

Open-ended probe technique measures the dielectric parameters from the phase and amplitude of the reflected signal at the end of an open-ended coaxial line that is in contact with the sample to be measured. Care should be taken to avoid any air gap between the probe and the sample. This technique is convenient for measuring dielectric properties in wide range of frequencies on non-porous material of relatively high loss factor, which includes most of the food materials.

Cavity ferturbation method

Cavity perturbation technique measures the dielectric constant and loss factor on the basis of the changes in the frequency and the width of transmission characteristics, when a sample is inserted into a tuned resonant cavity. Cavity perturbation technique is very sensitive and accurate for materials with loss factor less than one. This technique is suitable for non-porous materials.

6

Food Producers and Processors

Today consumer food choice is more complex than ever before. Consumers have developed more dynamic, complex and differentiated demands. These changes in consumer behaviour, reinforced by changes in the retailing sector, provide both threats and opportunities for the food sector. On the one hand, they offer new opportunities for adding value and differentiating products, which can lead to less price competition, strong consumer preferences, brand equity, better negotiating power facing retailers and higher margins. On the other hand, doing this successfully requires competencies, which many actors in the food sector have only to a limited degree, and in many cases it requires new forms of cooperation between the actors in the food chain.

Many food processors have been claiming that food consumers in the developed countries have become more difficult to understand and to predict. Some even claim that consumers act irrationally or even at random when choosing food products. We don't need to go that far, though - the complexity of consumer food choice can be viewed as a result of increasing differentiation of the food products to choose between on the one side, and increasing dynamics, complexity and heterogeneity of consumer demand on the other side. It is a complexity which is amenable to analysis, and which can be turned into a source of competitive advantage for food producers.

We can still assume that consumer food choice is influenced by food products' prices and quality and consumers' income. But what consumers regard as 'quality' has undergone considerable change during the past decades. We can approximate today's consumer food quality perception by distinguishing four groups of quality attributes for food products: sensory attributes, health attributes, process attributes, and convenience attributes. Sensory attributes refer to the classical food quality aspects taste, appearance, and smell, with taste as the dominant aspect.

Taste is an experience quality that can be evaluated only after the purchase, and consumers use a host of market signals, like brand, price, and quality labels, in trying to predict the taste experience. Health has been of increasing importance for consumer food choice for the last 50 years or so, and today analyses of consumer food quality perception many times indicate that health and sensory considerations have about equal weight. Health-related qualities are quite different from sensory qualities, though, in that they are, for the consumer, invisible.

While consumers have learned that there is a link between eating and health, they don't expect the consumption of a particular product on a particular occasion to have a health implication that they can experience. Many health effects of food are of a rather abstract nature - like the risk of particular diseases being reduced by a certain percentage - and thus do not lead to consequences that are readily accessible to experience. Health as a choice criterion for food is thus a question of communication and of the interpretation of various signals.

Messages about the healthiness of various types of food in the past have been conflicting, and consumers have constructed their own subjective theories of healthiness of food products. These theories depend on the learning history of the consumer and are thus individually different, but there are some recurring themes like that industrial production is less healthy than craftsmanship, that additives are unhealthy, that fat is bad and that vegetables are good. More recently, food manufacturers have used the health criterion more proactively in their product development through the development of functional foods, i.e., food products which have an added positive health benefit. Since the health benefit is still invisible for the consumer, the question of which health claims are allowed in the marketing of such products has become a major issue in the public debate.

Even though the type of health claim of course will have an impact on consumer food choice, it should be noted that such health claims, as all other relevant information, will affect consumer food choice not as such, but as interpreted based on consumers' subjective food health theories. Process attributes relate to consumers' interest in the way a food product has been produced, even when this has no analysable impact on the final food product. Consumers pay premiums for organic products which look and taste the same, for products which were produced with due concern for animal welfare and/or environmental considerations, and products which are guaranteed GMO-free.

Consumer interest in convenience has been rising quite considerably over the past few decades. Convenience is here defined as aspects of the product which save time or energy throughout the private household's meal production chain, i.e., during shopping, storage, preparation, eating, and disposal. In a wider sense, convenience relates to home-meal replacement alternatives like eating out. Increased demand for convenience has often been related to family demographics like both adult household members' participation in the work force, but newer research indicates that demand for convenience is best explained by households' subjective perceptions of their resource constraints.

Figure 1 shows results from surveys carried out in several European countries at two points in time, which allow looking at the development in the importance of the various food quality dimensions. In France, Germany and the UK we see the high importance of taste and health, but we also see that the convenience dimension is the only one that was on the rise in the period in question. While this bundle of quality attributes with an impact on consumer food choice already indicates considerable complexity of consumer choices, there are three additional factors complicating the issue. These relate to the existence of subjective trade-offs, cultural differences, and different consumer segments.

Subjective trade-offs relate to the phenomenon that consumers may be interested in all four types of quality attributes, but that they at the same time believe them to be, at least partly, incompatible. A product attribute like high fat in a dairy product may be regarded as an indicator of both superior taste and inferior health, as the example in figure 2 shows. Organics may be desired as a form of production but may at the same time be perceived as less convenient.

Convenience products with a high degree of processing may be regarded as undesirable in terms of their industrial way of production. These trade-off may be resolved by consumers in different ways under different situations, depending on the in that situation dominant buying motives.

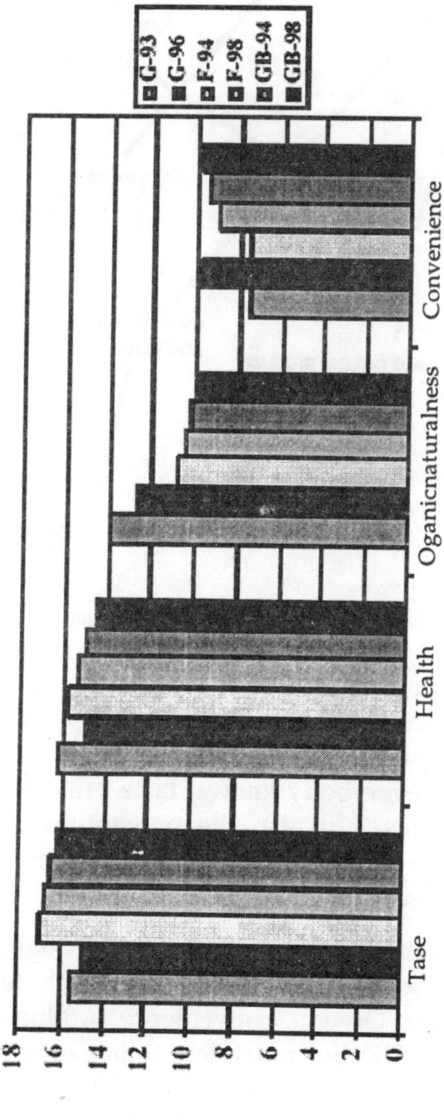

Figure 1. Importance of four major quality dimensions of food

Figure 2. Associations to 'high fat content' in a yoghurt product

It is widely acknowledged that food choice is subject to cultural differences. These differences will affect how the four types of quality attributes mentioned above affect food choice. As everybody knows, taste preferences differ between countries, but also the perception of what is healthy, what is convenient, and which types of production are acceptable may differ.

Animal welfare is a very desirable quality of the production process in the UK, but much less so in Southern Europe. Organic convenience products may be acceptable in the UK, but much less so in Germany, where organic production is mentally linked to low levels of processing. Acceptance of functional food products is higher in Finland, where biotechnology has a good reputation in the public discussion, but less so in Denmark, where consumers are sceptical about high-tech applications in food production, as the example in figure 3 shows.

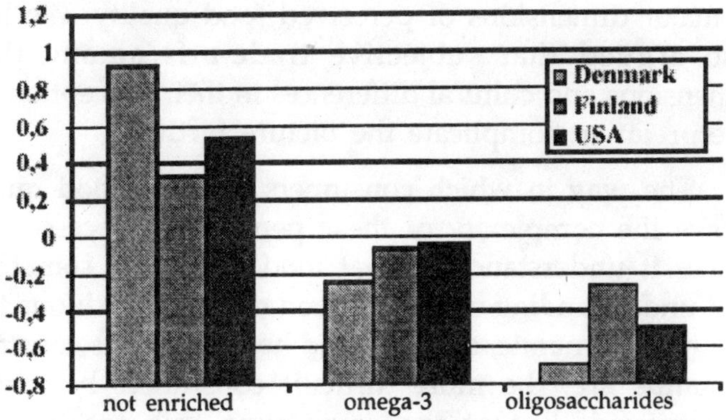

Figure 3. Predicted market shares for functional food product concepts by country and type of enrichment

Reanalysed from data from Bech-Larsen & Grunert, in press, based on conjoint analysis with juice and spread as carrier. Finally, even within countries, consumers may differ considerably in their preferences for food. In all countries we find consumers who are not very interested in food, and which put low emphasis on food quality issues. Likewise, there are consumers for whom food and eating is a source of stability and safety in their lives, and who therefore put high emphasis on traditional food qualities, known products and production methods, and non-innovative forms of preparing meals.

There are consumers with a very rational approach to food choice and others with an innovation-oriented, social and adventurous approach. Between these consumer segments, the weighting of the various quality attributes will differ, but also their interpretation.

That changes in consumer food choice can be fruitfully analysed as a process where the way in which consumers perceive quality in a food product has become more multidimensional. We have pointed at sensory attributes, health, convenience and process characteristics as major dimensions of perceived food quality. We have also argued that subjective trade-offs among these dimensions and cultural differences in their perception and interpretation complicate the picture further.

The way in which consumers perceive food quality shows the complexity of these perceptual processes and makes it understandable that food producers sometimes find understanding consumer food choice an arduous task. The more dynamic, complex and heterogeneous consumer demands are, the more difficult consumers become to understand. But at the same time this creates new opportunities for food producers for adding value and differentiating products, which we will come back to later.

Structural Changes in Retailing

The structural changes in retailing in Europe and the growth of retail private labels have been extensively documented. Private label shares of food sales for a number of European countries are shown in figure 4. It shows that there is considerable variation across Europe, with Switzerland and the UK topping the list. Eastern Europe is not shown in the table, but private label penetration in Eastern Europe is currently low. No matter what the current level of private label penetration, though, most retailers seem to have plans to increase it.

Several reasons have been suggested for why retailers try to increase the share of private label sales, including higher margins for private labels and better negotiating power for retailers facing national brand manufacturers. The aspect we want to concentrate on here is the changing role of retailers in the overall food value chain. The concentration process in the retail sector has changed both the power balance in the food chain and the competition among retailers. Retailers have adopted more proactive marketing strategies, where they try to achieve customer loyalty not only by parameters like service, location and store layout, but also by having more influence on the overall value creation process in the food chain.

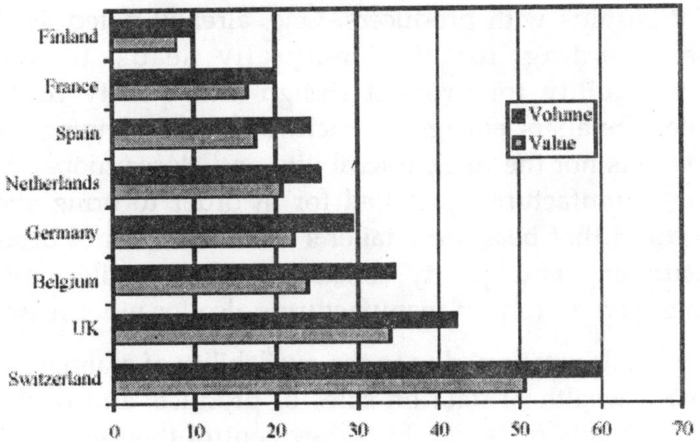

Figure 4: Shares of private lables in food by country From KPMG, 2000.

Private labels can be viewed as a major instrument for actively shaping the kinds of products consumers can choose among in the store, as these are the only ones where retailers can be sure that consumers cannot find them in competing stores as well. Consequently, there has been a trend towards the development of high quality, differentiated private label products, and the quality of these products has been shown to be a major determinant of consumers' decision to buy them.

These changes in retail structure and strategy have implications for retailer choice of suppliers. Branding, for consumers, has the functions of reducing purchase risk and of information costs. When retailers take over the branding function, they have to make sure that their brands also can fulfil these functions, if their brands are to be successful with consumers. We can therefore expect that the increased prominence of private labels also has an impact on the type of criteria retail purchasers apply when selecting suppliers and products.

There are several plausible explanations for retailers' increased interest in traceability and long-term relationships with producers. One, already noted, is that the branding function naturally leads to more responsibility for product design and quality control. Unless retailers engage in backward vertical integration (which is not the rule), traceability and closer cooperation with manufacturers is called for in order to bring about products that bear the retailer's mark in terms of design, positioning and quality consistency. But retailers many times have to rely on manufacturers also for other reasons.

While retailers, due to the availability of scanner data, have a wealth of data on sales of products and of those determinants of sales which they control themselves (like price promotions and shelf allocation), many retailers have little or no knowledge on determinants of buying behaviour on the consumer side. Manufacturers, which concentrate on a more narrow range of products than retailers, have many times a better understanding of consumer demands than retailers, and retailers may want to draw on this expertise when developing private label products.

Competitive Advantage of Food Producers

In consumer behaviour and in retailing, create both threats and opportunities for food producers and thus their opportunities for achieving a competitive advantage on international food markets. The changes in consumer behaviour described above imply changes in the way consumers perceive value in food products. While earlier on food products possessing basic qualities in terms of taste and nutrition may have been perceived as good value by consumers, the trends described above imply that the perception of high value will be linked to a broader array of factors, covering the four groups of quality attributes discussed earlier, namely sensory, health, process and convenience quality.

At the same time, the described individual, segmentspecific and cultural differences will imply that the process by which a consumer forms value perceptions when buying food will vary widely, especially from the viewpoint of an internationally operating food producer. These changes in the ways in which food consumers perceive value in food products therefore require food producers to possess not only competencies in producing food products meeting these dynamic, complex and heterogeneous demands, but also competencies in being able to understand these diverse ways in which consumers see value in food products.

The SPP framework is a dyadic model, and we have interpreted it as describing a dyad consisting of a food producer and consumers (end users). We may use the same framework to look at the dyad consisting of a food producer and its retail customers. Also retail buyers perceive, to varying degrees, value in a supplier and its products, as described above. The extent of perceived value will, among other factors, depend on the extent the retail purchaser believes that the supplier will be able to support the strategy of the retailer and the extent to which the products supplied will, in turn, be perceived as good value by the retailer's customers, i.e., consumers. We can thus extend the dyadic SPP framework to a value chain framework, where each member of the value chain perceives value in the supplies sourced from the preceding link, but where this value perception will, at least partly, be determined by assumptions on the way value is perceived by the chain members further down the chain.

From this perspective, we can interpret the trends in retailing in their implications for the joint value creation by producers and retailers. Also retailers are confronted with the consumer trends described earlier and have to address dynamic, complex and heterogeneous consumer demands. In addition, they want to be more than just a platform for producers' attempts to meet these demands; they want to take an active part in the process of filling them. In fulfilling this active role, they need a considerably higher degree of cooperation with food producers than earlier.

From the food producer's viewpoint, the developments in the retailing sector thus not only mean that retail purchasers' requirements have changed. They mean most notably that there is a new role for food producers in joint end user value creation together with retailers. This requires competencies not only in being able to produce those products retailers want, it also requires competencies with regard to the management of relationships between producers and retailers. The trends described above thus have changed the ways in which a food producer can attain a competitive advantage.

There are new ways of creating value in the eyes of the consumer. There are new ways of creating value in the eyes of the retailer. And there are new ways of creating value for consumers by cooperation with retailers.

Market-related Competencies

In order to turn the opportunities which consumer and retail trends offer into competitive advantage, food producers thus need to develop certain competencies. We can here roughly distinguish between production-related and market-related competencies, where productionrelated competencies deal with how to produce something and market-related competencies with finding out what to produce and how to market it. Market-related competencies allow food producers to be market-oriented.

Market orientation, often defined as "the organisationwide generation of market intelligence, pertaining to current and future customer needs, dissemination of the intelligence across departments, and organisation-wide responsiveness to it", has been identified as a major driver of competitive advantage, also in the food sector. In addition, market-related competencies have been shown to posses many of those characteristics which make competencies both rare, valuable and difficult to imitate, i.e., make them sources of sustainable competitive advantage.

Thus, market-related competencies typically are socially complex, causally ambiguous, involve interaction among humans, and involve a good deal of tacit knowledge, and their effectiveness may increase over time. An emphasis on market orientation and market-related competencies will typically also have implications for production-related competencies. Since market orientation will have its major positive impact on competitive advantage in situations where customer demands are dynamic, complex and heterogeneous, being market-oriented will result in the adaptation of existing and the development of new products.

A higher degree of market orientation will therefore usually result in a demand for more flexibility with regard to the production-related competencies. This goes both for flexibility in using available resources in alternative operations, typically required when products are differentiated, and for flexibility in identifying, configuring, and deploying chains of resources for new uses, typically required in the context of new product development. Decision-makers in food companies are clearly aware of the importance of market-related competencies, but also of the need for further development of these competencies.

A 'deficit index' was computed by subtracting the proficiency rating from the importance rating, so that higher values indicate a larger discrepancy between perceived importance and own proficiency. It is clearly seen that market-related competence elements top the list, related both to obtaining market intelligence and to using it in the organisation, especially in the context of new product development. Efficiency- and production-related competence elements are further down the list. In another study, groups of industry leaders were confronted with information on relevant market trends and were then asked, using a reversed laddering procedure, to translate these into relevant competencies to be developed in their organisations in order to be able to meet these market trends.

The development in consumer demands opens up for new possibilities for adding value and differentiating products, but since there are so many options for adding value and differentiation, a close link to understanding consumers becomes crucial. This should not be confounded with asking consumers which kind of products they want. Experience in product development shows that, when asked directly, consumers can mostly only come up with suggestions for improvements of existing products.

Successful consumer understanding implies understanding the mechanisms underlying consumer food choice, the trends in the development of major purchase motives, the role of situational factors in food choice etc. Because of segment-specific and cultural differences, as discussed above, such understanding will not travel easily from one market to another, and the more a food producer aims to build a competitive advantage based on high value-added products, the more it may become necessary to concentrate on a few markets, where the necessary degree of consumer understanding can be achieved.

Also, such knowledge becomes useful for food producers only to the extent it can be turned into applications in the organisation, i.e., is fed into the company's organisational learning process. Consumer understanding is thus not a competence that can be outsourced; in a food producing company that aims at exploiting the opportunities of trends in consumer food choice, the ability to analyse and understand consumers becomes a core competence. Consumer understanding can lead to the attainment of competitive advantage to the extent it contributes to the successful development of new products.

Developing new products is a risky and difficult matter, as witnessed by the high failure rates of new product introductions. Decades of research on success factors in new product development, some of which has also been done in the food industry, have resulted in a considerable body of good advice with regard to the management and organisation of product development, and with regard to the importance of linking the product development process to the market, by providing market input throughout the product development process from idea generation through concept testing to the development of prototypes.

In developing modern food products matching modern consumer trends, it should especially be noted that many of the unique qualities of these products, like health effects and methods of production, are invisible to the consumer and therefore have to be communicated. In these cases, the contingent development of the physical product and the communication about the product becomes an important competence element to develop. Inputs from consumer analysis, which can provide guidance for the development of both the physical product and the communication, can be a way of furthering this contingency.

Downstream, managing relationships with retailers by responding to retailers' call for traceability and long-term relationships, especially in the area of private label products, is a way of responding to the changes in retailers' role. Upstream, managing relationships to suppliers becomes more important whenever new ways of differentiation and value adding require changes already in the raw materials or primary supplies.

Differentiation in primary production or early in the value chain requires the resulting products to be kept separate from the rest throughout the rest of the value chain. This by itself increases transaction costs; in addition, the asset specificity associated with investing in production methods and products for more specific applications and target groups will entail that the optimal governance structure in many cases changes from spot markets to long-term relationship economics.

Competence in developing new products will be important in both cases, but the implementation of this competence will differ: when the producer enters a long-term partnership with retailers, much of the product development process will be a collaborative effort, with frequent communication with the retail partner.

7

Processed Foods

Increasing agricultural exports from developing countries can have important benefits, as agriculture remains the major source of livelihood and food security for large sections of the population. If SPS (Sanitary and Phytosanitary Measures) measures become too stringent, the resulting contraction in agricultural exports would have a significant negative impact on such communities. Even the WTO has recognised this, and it has highlighted the need for consumers to compensate producers who would be adversely affected in this manner. It has also pointed out that the removal or relaxation of overly stringent SPS regulations could generate welfare gains to consumers. Obviously, in the global context, the majority of producers fall in the developing country category and may not be able to effectively access the developed country markets if restrictions are very severe.

The significance of SPS measures in today's agricultural commodities trade can be gained from the number of notifications circulated since the agreement came into force in 1995. Nearly 2630 notifications have been circulated from 1995 to the end of 2001. During this period, the USA with 526 submitted notifications and Mexico with 175 were the most proactive trading countries in as far as SPS measures are concerned. It is reported that USA notified the largest number of SPS measures in 2001. In fact, it is estimated that OECD countries have submitted two out of every three notifications. Although, these notifications are submitted in conformity with the transparency clause (Article 7) of the SPS Agreement, these figures indicate the legislative complexities of SPS policies in the countries that dominate international trade in agricultural commodities. The fresh flow of notifications indicates that higher and more stringent SPS standards (that go beyond the international standards are being imposed by individual countries to protect human as well as plant and animal health.

The proliferation of higher and stringent food safety standards drawdeveloping countries into an unknown territory of international trade for which they are ill equipped and unprepared. Interestingly, some of the more buoyant export commodity groups for India, namely, coffee, spices, fresh fruits and vegetables and pulses, will now have to reckon with more stringent certification and labelling provisions notified by the developed countries. Producing for export also requires sensitivity to the changing demand scenario in the world market, which, in recent times, is greatly influenced by food safety standards.

Therefore, the traditional wisdom that attributes the growth and importance of agricultural exports to lower capital per unit of output, shorter gestation period and strong backward and forward linkages requires a closer examination. These linkages are of paramount importance for several reasons. For example, value added processed food exports require two necessary conditions: (i) a threshold base of manufacturing and other infrastructure facilities and, (ii) a reasonable level of marketable surplus in those primary products that enter as crucial inputs into food processing (industrial activity) chain. The food processing technologies, admittedly, range across a wide spectrum but a successful and viable business enterprise must be able to harness economies of scale.

The economics of the processing plant depends critically on having a large volume of raw material supply of adequate quality. The supply chain, being biological in nature, often requires extra effort for timetemperature domain management when integrating with industrial applications. In processing, principal raw materials and by-products to enhance value-addition, the food quality dimensions must not be lost sight of. In as much as microorganism contamination along with additives is critically important in determining compliance with food safety regulations, FATTOM management in the manufactured foods becomes significant.

A clear understanding oft the dynamics of FATTOM can be used to regulate microorganism growth in the processed foods. Developed countries, may have a distinct edge over the developing countries in the manufacturing domain because they have better access to processing technology and packaging innovations. These initiatives coupled with a distribution network do cater to the changing dietary patterns across space and age. Hence, the developed countries accounted for around three fourth of world processed food exports in 1999, though the developing country share had increased somewhat after 1995.

The contrast can be well illustrated by the fact that India, though the largest producer of bovine milk in the world during 1999, had a mere 0.5 per cent share in the processed food exports. Even the marginal gains by the developing countries in processed foods are confined to a few categories; in 1999, for example, these were visible only in Fish Products, Vegetable Oils and preserved Fruits and Vegetables. This explains to a considerable extent why developed countries issue most SPS notifications. These aspects must be taken into account when considering the post-WTO international trading system and the SPS regime.

Growth of Food Industry

An in-depth analysis of the evolution of the food processing industry in India requires time series data with a comparable data set and commodity basket composition. Such an analysis can throw better light on the dynamism of the processing segment in food exports and permit comparison with experiences elsewhere in the world. However, it should be recognised at the outset that the food processing industry per se is a distant cousin of the manufacturing industrial structure/sector. It is dependent, among other factors, on the agricultural sector's output as well as on consumers' dietary preferences.

The domestic situation on both these counts did not encourage rapid growth of the food processing industry. While the international scenario in that respect needs to be properly situated and carefully considered, the critical factor has been that the myriad dimensions of food security management in India ipso facto laid greater emphasis on enhancing food production rather on food processing.

An immediate consequence of the trade liberalisation initiated during the early 1990s can be seen in the growth rate of processed food exports. A preliminary examination of the export performance of processed food products does suggest a priori that the incidence of SPS regulations impacting on India has fallen mainly on food products that have undergone a higher degree of processing. We can draw out some broad contours of the evolutionary path of processed food exports from a careful scrutiny of the data on the share of different food products. The share of the two categories, meat and meat preparations(3111), and fruit and vegetable preparations (3113), do not show much change over the period 1981-1999. However, after 1991, year-to-year variations are more pronounced in 3111 in comparison to 3113 and these may be related to the role of SPS regulations in different export markets. If we consider the post-1991 period and take out the share of Tobacco manufacturing (3140), the total share of the remaining commodity groups indicates the growing importance of grain mill products (3116).

Again, the research issue would be to assess the impact (if any) of the SPS regulations. Interestingly, the share of food classification 3121 over the same period shows a decline from about 30 per cent to about 20 per cent, almost changing places with 3116. The history of agricultural exports from India is a story of considerable year to year variations, but processed food exports are of different kinds and buoyant exports to matured market would require more intensively processed food products. It is in these markets that stricter and higher than international reference standards are relevant in market access issues.

These figures suggest that in the post-economic reforms period since 1991, the indirect effect of trade liberalisation, exchange rate adjustment and relaxation of restrictions on agricultural exports have been positive and significant. The four selected processed food products (3114, 3115, 3116 and 3121) account for about 88 per cent of the total export earnings during 1999, that is 3-percentage points higher than the share in 1995. Interestingly, grain-milling products that comprise intermediary processed foods show significant growth during this period.

The Japanese stipulation of Vapour Heat Treatment (VHT) of fruits is yet another instance of SPS becoming the key non-tariff barrier. The technological upgrading to comply with VHT protocol is a story of investment of time and money for at least five years. This is in spite of the fact that success at the end is not assured. The introduction of a regulation by EU prescribing very low levels of Octratoxin-A (OTA) in coffee; method and sensitivity of estimating pesticide residue in vegetables, fruits, honey, etc. appear quite unreasonable. Another instance is the EU demand for the residue monitoring plans for the previous years in association with that for succeeding years. This stipulation will definitely deny access to the EU market for Indian agro products.

The labelling stipulation in the importing country language too is a costly proposition keeping out Indian exporters. As a consequence of all these measures, costs are imposed on the exporting country without any expected commensurate return. To sum-up, the issues of food safety regulations for a country of India's size with a wide spectrum of agro-climatic dimensions require a detailed examination in a logical framework of different processes.

The need for such an examination can be highlighted if one could visualise a food market output and export matrix or the volume-value matrix with the respective shares for the country. The SPS relevant exports have strong backward and forward linkages with deep ramifications. Finally, processed food product lines depend on a host of players in both exporting and importing countries. India has an elaborate system of quality inspection and certification before any product is exported. In recent times, more rigour has been brought into this process and the domestic system is evolving in response to the reported number of rejections of exportable commodities. The Export Inspection Council (EIC) is the apex-designated agency that is charged with this responsibility. For brevity, SPS compliant exports are facilitated by the EIC, which imposes a system of three types of inspection and certification, namely consignment-wise inspection; inprocess quality control and a food safety management system based certification.

However, for brevity we may recall that under the Consignment Wise Inspection (CWI), each export consignment is inspected and tested by the recognised inspection agencies. Samples are drawn on the basis of statistical sampling plans, inspected and tested for verifying the conformity of products to the prescribed standards. The in-Process Quality Control (IPQC) system lays emphasis on the responsibility of manufacturers/ processors in ensuring consistency in quality during all stages of production by adopting quality control drills and exercising control on raw materials and bought-out components, manufacturing process, packing and final testing.

Manufacturing and processing units, adjudged as having adequate levels of quality control in all these areas, are approved by EIC based on the assessments. Units approved under this system are eligible to get certificate of export worthiness without further verification of the quality of the out going consignments by EIC and random spot check of the consignments are carried from time to time. Under the simplified inspection procedure, such units have been given the option, either to issue certificate of inspection of export worthiness on their own or to obtain certificate of inspection from EIC.

In view of growing concern the world over regarding health and safety parameters of food items being imported, international standards on Food Safety Management Systems like HACCP/GMP/GHP have been developed. Based on such standards, which are being prescribed by several of India's trading partners such as European

Union, etc. EIC has introduced certification of product quality integrated with the systems approach. Currently, Fish & Fishery Products, Egg Products and Milk Products are being certified under the above system. Agricultural exports in the post-WTO period are dependent on import procedures in the destination country. The US is probably the only country, which provides information on the detention of shipments based on a pre-inspection basis.

Product Lines

Processed food products are gaining in importance in the export basket of India. There appear to be two key issues relevant for SPS compliant exports: (a) the pre-shipment inspection and certification by the exporting country and, (b) the import procedures and detention in the destination countries. Since, there are no uniform or homogenous product lines, there are wide variations in relevant food safety norms and over time, food safety standards have tended to become more complex and vary substantially amongst countries. The selected product lines are Poultry products; Marine products; Mango Pulp, Peanuts and Mushrooms, which have contributed substantially to India's processed food exports in recent years.

Poultry products

It may be appropriate to recall that two years back, India was de-listed from the EU's list of approved countries for import of egg powder into EU because of its non-submission of a Residue Monitoring Plan (RMP). It has been the strategy of EU countries to introduce newer, and stricter residue limits every time a need arises to restrict imports from developing countries like India. The issue of residue limits and the Residue Monitoring Plan itself has indeed been used as an SPS measure very strongly by developed countries like EU as well as the USA. On the other hand, India also suffers since no domestic agency took the responsibility of preparing the Residue Monitoring Plan for animal products including egg powder, and the matter was thrown from one Ministry/ department to other.

A second type of SPS measure employed by the developed countries is the granting of equivalency to countries like India. EU and USDA have not made the effort to grant equivalency to Indian standards for egg powder. Even after four years of having submitted the list of plants to be notified by EU, they have not constituted a commission to inspect these plants in India, notify them and grant them equivalency. A third type of SPS measure used by some countries is certification of labs.

The test certificates issued by Indian laboratories are invariably not accepted in EU and other developed countries as these labs are not accredited to the labs of developed countries, even though the Indian labs follow the same testing methods and protocol for testing the samples. The US importation rules, clearly state that importation of poultry and poultry meat products from other countries can be on various grounds, which favour domestic producers. The Food Safety and Inspection Service (FSIS) can suspend the eligibility of another country if it feels that an emerging sanitary measure is to be implemented to address a hazard that is so severe that no product can enter from a foreign establishment until a control is in place.

In a second situation, if the other country does not provide satisfactory documentation of equivalent sanitary measure or if the FSIS audit reveals that the exporting country is not implementing a public health sanitary measure in the manner that the FSIS determined it to be equivalent, they can permanently stop eligibility of that country for export. They can further take action against a particular country if they feel that their products are adulterated or misbranded on on-site audit or because of Port of Entry re-inspection, etc. These are SPS measures in different garbs, which are or can be used for stopping exports from developing countries like India.

The poultry industry consists of producers of both layers and broilers. The production of eggs, and chicken meat represent different stages, starting from the Great Grand Parents or Purelines, which are followed by the next generation of Grand Parents and Parent Breeding Farm. Up to this stage, the science involved is pure genetics followed by sound principles of poultry management. Hatchery is the hub area both of broilers and layers, where sanitation and hygiene play a very major and critical role. This is one area where, unless a lot of care is taken, the chicks produced from incubators and hatcheries can be infected with different diseases. These will not only affect the health of the birds, but can also create food safety problems for the consumers. There is a very close inter-relationship between such stages and each link has to be protected from contamination.

The production chain can be vertically integrated or several independent firms, depending on their respective core competencies, can undertake production at each stage (producing Purelines, Great Grand Parents, and Parent Stock). Similarly, independent firms can also produce day-old chicks in their hatcheries, which can be established by a parent-breeding farm. Food, in the form of broiler chicken and eggs, can be produced by the integrator himself or it can be purchased from outside sources (e. g. from farmers). In India, vertical integration has not taken root very strongly.

There are only a few companies, such as the VH Group of Companies, that are involved in all the activities of the supply chain in a typical integrated operation. Independent producers of Grand Parents and Parents, and a large number of hatchery operators run most of the poultry operations. Similarly, integrators produce poultry feed, with many companies like Hindustan Levers, Godrej, Uttara Poultry Feeds, Poshak, etc., involved in feed production. There are a large number of commercial farmers producing layers and broilers. Further processing of chicken and eggs is still at a nascent stage. There are about half a dozen egg processing plants, of which three are currently in operation and two of them are HACCP compliant, meeting the international standards.

During 1996, all 6 plants were exporting egg products to EU and other developed countries. But, with the imposition of the new residue limits for pesticides, and the requirement for submission and execution of RMP by India as country, egg powder exports declined sharply. According to industry sources, India has lost an export market of egg powder of more than Rs 100 crores during last three years. This greatly lowered capacity utilisation of almost all units, and led to the closure of 3 units. After 3 plants gained accreditation for higher standards, Egg Powder exports started picking up again in 2002. To become HACCP compliant, each unit had to invest around Rs 1. 5 to 2. 00 crores, which is around 5 per cent of the total investment cost.

The new plants need an in-built lab to check the validity of compliance, and various inputs and skilled labour, all of which raise operating costs. With no domestic market for egg powder, the existence of nonoperating units, low capacity utilisation of operational units and their higher operating costs adds up to a huge burden on the industry. A few firms, whose plants meet international standards and who also supply to multi-national food chains like Domino's, Pizza Hut, KFC and TGI Fridays, do further processing. About halfa-dozen plants in operation also do whole bird slaughter and processing.

About 97 per cent of the chicken is still sold live in 'Mandies' (wholesale markets). In these 'mandies' or market yards like Gazipur of Delhi, more than 2,00,000 birds are traded everyday.

The broilers are still sold live, and that too on a score basis (20 numbers make one score). Some layer farms have started egg grading, washing and packaging for export purposes. However the majority of the eggs are still sold in small numbers by small and marginal operators. This complex situation, with many players in the chain, raises difficult challenges in terms of moving to a vertically integrated system. Food safety, however, becomes a major problem as there are so many handlers doing different jobs, and many a time they may not even be aware of the food safety requirements. Many companies who operate at the primary breeding stage where the rest of the chain is integrated and control is centralised may be in an advantageous position in terms of food safety, relative to those that are not integrated.

In countries like India, where the live market still dominates in the chicken industry, with a large number of butchers and processors, standards of food safety are difficult to meet. Further, there are no documented guidelines - either voluntary or mandatory - that can be enforced. To achieve an acceptable level of confidence in food safety, appropriate steps pertaining to risk assessment need to be undertaken.

Marine products

Marine products have long been the most buoyant among Indian export lines, following the imposition of stringent quality controls for marine products after the SPS regulations came into force. The demand for stringent and high hygienic standards in the production and processing facilities greatly increased, after the stipulation of Hazard

Analysis Critical Control Point (HACCP) by United States Food and Drug Authority (USFDA), ISO9000 and other European Community directives, and the EC ban on Indian marine products in 1997. The impact is most clearly felt in the production of individually quick frozen (IQF) and other value added frozen items for export to the major overseas markets.

Marine products, on account of their health attributes and high unit value, are claimed to be one of the fastest moving commodities in world markets. The world market for seafood is reported to have doubled between 1987/88-1997/98, reaching the US$49. 32 billion mark, of which India had a 2. 4 per cent share. India has depended on Shrimps as the major export product while the world market is fast changing the composition of the seafood basket. Consequently, Indian exporters are attempting to penetrate into the new markets of Europe and South East Asia. But, in value terms, frozen Shrimp remains the largest marine food export product accounting for about 69. 50 per cent in the total value of marine product exports from India during 2001-02, showing a slight decline from the over 71 per cent share in 1995.

The Indian marine products exports are driven primarily by the Japanese and the European Union (EU) markets. For example, exports to Japan increased from US$251. 49 million in 1987/88 to US$641. 68 million in 1997/98, whereas during the same period exports to EU grew from US$60. 76 million to US$113. 80 million. Since then, marine products exports have grown further due to rapid supply expansion through shrimp farming and the introduction of several resource-specific vessels to enhance marine fish landings. In 2001-02, Japan (30. 56%), USA (23. 9%) and EU (19. 31%) together accounted for about 74 per cent (by value) of India's marine product exports, while they had accounted for about 71. 4 per cent in 2000-01.

The value of marine product exports to Japan declined by almost a 32 per cent from US$562. 75 million in 2000-01 to US$383. 07 million in 2001-02. On the other hand, exports to USA and EU markets increased during this period by almost 25 per cent. Given the dominance of these markets in total shrimp exports from India, the importance of quality control becomes critical. Processed marine products differ widely, and deteriorate rapidly in tropical conditions. The EC directive no. 91/493, dated 22 July, 1991 that came into force from 1 January, 1993 made it mandatory to comply with specified health conditions for the production and placement of these products in the EU markets, and in August 1997, it banned fishery products from India. This precipitative action was justified by the EC on three primary grounds:

— Serious deficiencies with regard to infrastructure and hygiene in the fish processing establishments;

— Potentially high risk for public health given the conditions of production and processing of fisheries products;

— Contaminated by micro-organisms, which may constitute a hazard to human health.

On the other hand, USFDA came out with a new version of the HACCP based inspection proposal for seafood on 28 January, 1994. After a series of public audits and publication of the inspection procedure, the mandatory nature of USFDA was made clear. The Government of India responded to these developments by taking important steps to maintain the highest quality standards based on the health safety regulation requirements of the importing countries. The Seafood Exporters Association of India claims to have spent US$25 million on upgrading their facilities to meet the food safety regulations of the importing countries.

The EC standards are higher than the HACCP standards and compliance requires going through the process of gaining EC approval of plants. The EC approval involves, in essence, a series of domestic legislative and EC adjudicative steps. Though the domestic regulatory regime has been made compliant to WTO stipulation, ECs adjudicative role has been subjected to critical examination The adjudicative problems with quality compliance can be looked at from three broad perspectives:

1. Pre-processing, including Shrimp aquaculture and handling of raw Shrimp at various stages, such as harvesting, sorting, etc.
2. Processing, wherein water quality, source of water for ice making (62 tests to ensure water hygiene), infrastructure (size of the wash room) and transportation utilities come in for sanctions.
3. Post-processing, including testing, packaging and marketing activities.

A case study indicates that Indian exporters have to incur large costs if they want to access the upper quality market. This heavy cost burden involves both fixed and variable cost components. Given the small and medium size of most processing units, an additional fixed investment to the tune of Rs. 1-2 crores per unit for upgrading, often becomes prohibitively high in the absence of a well-timed and concessional loan facility. These plants, it may be recalled, are not exclusively dedicated plants, and with no assurance of market access even if this kind of large investments are made to upgrade infrastructure, many existing players will find it impossible to continue in the business.

Interestingly the overhead constituent of the variable cost component of compliance for a medium size plant here is estimated to go up by a factor of 5. According to

SEAI, the compliance cost for meeting the EC norms is 15-40 per cent of value for new units, with the cost being higher for existing units. It is estimated by MPEDA that only two units in three may be able to upgrade themselves to the EC norms, while the rest would be forced to close down and exit from the business. Thus, there are issues of sustainability in the marine export business. While, exporters have been pleading - so far unsuccessfully- for permission to move into deep-sea fishing, the government is seriously concerned about the fragile marine eco-system in the Shrimp habitats Aquaculture is one available alternative, and MPEDA is assisting it with provision of technical assistance.

The total area under Shrimp farming at the end of 2001-02 was estimated at 1,35,077 hectares. Of this, more than 50,000 hectares is devoted to traditional shrimp farming practices in the states of Kerala, West Bengal and Karnataka. The remainder is under scientific farming, with the active assistance of MPEDA. The area potentially suitable for Shrimp farming along the coast in India is estimated to be 1. 2 million hectares, of which only about 10 per cent is currently being utilised. There is therefore much scope for improving the production.

Environmental issues have emerged in aquaculture too, but these involve domestic environmental concerns rather than international sustainability issues. The concerns arose in view of the reported ecological and environmental effects of aquaculture in Southeast Asian countries. Experts, however, suggest that, as far as India is concerned, these concerns are misplaced, arguing that apprehensions that Shrimp farming causes degradation of the coastal zone lack any strong scientific backing. In fact, the setting up of aqua farms in the coastal zone has helped to protect the zone, as most of such units have taken care to construct proper granite bunds on the outer areas facing the sea coast.

In a way, these farms protect the coastal zone against sea-erosion during the monsoon. Aquaculture units are set up in fallow areas where land is inundated with saline or brackish water, and they do not encroach on traditional fishing or farming zones. Concerns have also been raised about the acute shortage of drinking water in the coastal areas and there have been suspicions that aquaculture may have contributed to this. As the Shrimp farms are mostly located in hard clay soils that have very little the seepage, this fear too appears to be misplaced. Unlike countries like Taiwan or the Philippines, India does not use groundwater for aqua-culture and as imported and costly seed is used, there are incentives to use resources carefully, and hence aquaculture tends to check environmental pollution and degradation.

Effluents from Shrimp farms are biodegradable, but intensive culture systems aimed at high levels of production per hectare can generate pollutants in the form of heavy metals (mercury, cadmium), pesticides and petroleum products. This has led the Government of Orissa to ban aqua-culture around the Chilka Lake. The solution to this problem is to discourage intensive culture systems and MPEDA, in its technical assistance programmes, recommends a sustainable farming system. In fact, aquaculture provides an environmental win-win situation in coastal Kerala, where rice and shrimp crops can be rotated on the same land. In fact, this has been traditionally practised in that area.

Aquaculture cannot be practiced during the monsoon season, and takes only three to four months. On the other hand, rice can be grown only during the monsoon. It is a fact that some aquaculture farmers have purchased land at a premium from traditional agriculture farmers, and to that extent there is a shift out of agriculture. This trend should be checked, at least in the interior regions, and this can be done by the states concerned through the Land Utilisation Act. The main environmental issues for aquaculture are, in fact, of a different kind. For example, degradation of aquaculture land due to pesticide residues discharged from agricultural land is threatening aquaculture activity.

Effluents from industrial belts along the coast may also contribute to such degradation. The fact that fish cannot survive in polluted water can assist policy makers, who can identify areas needing corrective measures by looking at the aquaculture units in the area. Clearly, it is a profitable business and quite labour intensive (650 man-days per hectare as compared to 50 man-days in traditional farming), so that it can generate considerable new employment. In Orissa, exporters claim that 8 per cent state government sales tax is a burden, which falls even on exports, thus weakening their competitive position in export markets. Value market share of Shrimp and its preparations from India declined following the 1997 ban by EU and the subsequent threat by USA relating to the Turtle Extruder Device (TED).

Although, Shrimp exports were 69. 50 percent of the total value of marine export lines during 2001-02, the unit value realisation remains low. This is due mainly to the high compliance cost of SPS measures that are estimated to increase operating expenses by a substantial factor. In the absence of any assurance of market access to developed countries, exporters are exploring other markets. Given India's long coastline and the availability of distinct marine life, especially the Tiger Shrimp, the scientific dimensions of the sustainability of aquaculture should be assessed from the view point of the economic viability of small and medium enterprises (SMEs).

SMEs enhance social welfare, and there is scientific knowledge that is enshrined in the indigenous system of marine husbandry. The future of an entire social group dependant on marine/fishery for their livelihood is determined by choices made by a large number of consumers, business houses and government officials. This is an issue that must be brought to the top of the trade agenda.

Peanuts

Peanut exports are handled by APEDA, and forms a major part of the commodity basket in the broad category "Other Processed Products". We intend to use Peanuts and Groundnuts interchangeably since the APEDA terminology depending on the HS classification uses the word Groundnuts. The export value of Groundnuts, declined to Rs. 250. 94 crores in 2001/02 from a high of Rs. 566. 3 crores in 1997/98. Peanuts comprised almost 40 per cent of "Other Processed Products" in 1993/94 and 1997/98, but accounted for only 14 percent by 2001/02.

In fact, the final export value in 2001-02 (Rs. 250. 94 crores) is around the value realized during 1995-96, a period when strict standards were not the major concern in trade. Be that as it may, the relative position of groundnut in the category "Other Processed Products" has sharply declined to 14. 1 per cent in 2001-02. Indian peanut exporters express the view that non-tariff barriers are placed on their exports in foreign markets to protect high cost domestic agricultural producers. They also face situations where they are compelled to make distress sales when foreign buyers are unable to accept their products because of non-compliance with some standards in the importer's market. Hence they feel that Indian exporters may have to depend primarily on their domestic market and, perhaps, to some extent, the SAARC region. The export data for 2001-02, in fact, reveal that about 75 per cent of the total groundnuts export value were realized from three destination markets

Some problems faced by these exporters appear to be genuine. For example, different testing procedures and conformity assessment standards are required in different markets. Though each test costs Rs. 6000, no one has communicated to the exporters why most of these tests are needed. Further, EU requires these tests only for imports from Egypt and India, while imports from countries such as the USA and Argentina are exempted from such tests. Another problem is that while there is no import duty on 50-kg bags, there is a duty on 5-kg bags, as the foreign markets want to discourage retail consignments. Exporters also face problems regarding genetically modified peanuts (GMOs).

Some years ago, a particular foreign market encouraged the use of GMOs; however, another market has now asked for an assurance that peanuts supplied are free of GMOs. The issue of aflatoxin presence in peanuts appears to be a major threat to peanut exports. The EU Commission in Brussels has specified tolerance limits for aflatoxin contamination in peanuts along with the appropriate testing methods. The proposed levels are 10 ppb (5ppb B1) for raw material and 4 ppb (2ppb B1) for consumer ready products. The new proposed sampling plan is similar to the Dutch Code, i. e., the analysis is to be done based on a 3-test Dutch code methodology from a randomly drawn 30 kgs sample.

The new procedure is much more rigorous than that currently in force, as if any of the 3 tests indicates that limits are exceeded, the lot will be rejected. These limits are not warranted on scientific grounds (as pointed out in submissions made by various agencies and governments to the SPS committee of WTO). Laboratory tests with small animals (such as touts and rats) which were fed highly contaminated feed (B1) on a daily basis indicated that aflatoxin could cause cancer of the liver. But there is, as yet, no clear evidence that that aflatoxins are carcinogenic in humans.

Even in major producer countries like Argentina, China, India, South Africa, U. S. A., Vietnam, etc. where peanut consumption is very high, particularly among the middle and lower income groups, there have been no findings or reports that aflatoxin in peanuts has increased cases of liver cancer. And peanuts are consumed in a very big way by all strata of society, especially the middle and lower class. Further, even if a shipment of peanuts is found to contain aflatoxin, this does not mean that all peanuts are contaminated since aflatoxin is concentrated in a small number of nuts.

Statistically, one would expect to find one contaminated nut in a sample of say, 5000 to 10,000 uncontaminated nuts. Experts have concluded that 75 per cent of the lots rejected under the proposed procedure would have aflatoxin levels below the established tolerance limit, i. e. they would be basically uncontaminated material. According to a Joint Expert Committee on Food Additives (JECFA) report, aflatoxin contamination of foodstuffs is very low among EU nations, and only a few people suffer from hepatitis B. At an estimated risk at 20 ppb, there will be only 0. 0041 cancer cases per 100,000 annually. Placing the risk at 10ppb, there will be only 0. 0039 cancer cases per 100,000 annually.

Hence the change of the standard from 20 ppb to 10 ppb would only reduce estimated cancer risk by approximately 2 cancer cases annually per 1 billion people. Thus, it seems improbable that there is any measurable risk differential between the two standards (20 and 10ppb) in populations with a low hepatitis B incidence (like in the EU countries). The JECFA had previously recommended that maximum permissible aflatoxin levels should be fixed as low as possible. But now, on the basis of further available data, it has modified its recommendation to a reduction in the intake as far "as is reasonably possible".

Further, it should be noted that the JECFA's risk estimates are based on data that made no allowance for the substantial reduction in aflatoxin contamination achieved by mechanical removal of the nut skins, and by the use of optical and electronic methods for sorting the nuts. The risk computations are thus based on incidence of aflatoxin contamination, which are no longer applicable. This should be taken into account when specifying the future EU tolerance limits. It should be noted here that the Codex Alimentarius Commission had proposed a maximum limit of 15 ppb.

The implementation of the EU Commission's proposals would endanger the export of Indian peanuts to EU member countries. The planned tolerance limits of 2 ppb aflatoxin B1 and 4 ppb total aflatoxin in finished products are so low that they would almost certainly cause insurmountable difficulties to potential exporters to EU countries and impose huge compliance costs. Producers within the EU itself would also suffer unreasonably from these regulations. Whereas the WHO is proposing a limit of 15 ppb for all aflatoxin, the EU Commission continues to insist on an upper limit of 10 ppb for the raw nuts, despite the fact that the aflatoxin content decreases during subsequent processing.

The European Snack Association's Nut Working Group has already expressed the industry's concern over the testing programme and analytical methodologies through CIAA (the European Food and Beverage Association). The American Peanut Council has submitted documents showing significant increase in costs and rejections as a result of the multi-sample system. The UK Ministry of Agriculture (MAFF) (UK is the largest consumer of peanuts in Europe,- and accounts for approximately 25 per cent of the peanuts imported into Europe) has already stated that the proposals imposed a higher burden than required by current UK regulations and could result in unacceptable costs to both industry and enforcement without any prospect of improved consumer safety.

Despite these protests, the revised draft of the sampling plan still recommends a multiple sampling system. It is evident that such a change will have very serious implications for the peanut industry. It is also noteworthy that this EU proposal possibly contravenes the GATT/WTO agreement, as it will erect artificial barriers and seriously discriminate against a number of producing countries, particularly those in the Third World and developing countries, including India.

Mango pulp

The mango pulp case, contrasts with the peanut case, and is illustrative in many ways. The value of mango pulp exports increased from a low of Rs. 26. 75 crores in 1991 to a high of Rs. 263. 85 crores in 2000/01, but declined to Rs. 241. 24 crores in the following year. There are only nine major exporters of mango pulp in the country. Sourcing is done primarily from the Chittoor District of Andhra Pradesh and Krishnagiri District of Tamil Nadu in SouthIndia. APEDA has taken important export promotion measures for Mango pulp. APEDA implemented HACCP with partial financial assistance from the Ministry of Food Processing Industries.

During 1997-98, 12 processing units in Chittoor District were taken up. Subsequently, 12 units in the Krishnagiri District of Tamil Nadu have been taken up, with an investment of about Rs. 3. 5 million. The compliance costs for implementing HACCP would have been prohibitive for the exporting firms, had APEDA not come to their rescue with both financial and technical assistance. All the participating units in the Chittoor District have implemented HACCP. Five units were assessed and certified by International Standards Certification (ISC) South Asia Pvt. Ltd. during the 1998 mango season. Six units of Chittoor District and 6 units of Krishnagiri District were assessed during the 1999 mango season.

The National Sanitation Foundation (NSF) has recommended all the participating units of Chittoor District for certification after the certification audit. In the case of units in Krishnagiri District, Quality Assurance Service (Australia) has carried out certification audits of 6 and all of these have been recommended for certification. However, small units have not been able to benefit from APEDA's efforts as problems in applying HACCP at the farm level occur given the nature of farms and practices in India. The quality norms under the Prevention of Food Adulteration Act (PFA) of India do not fully match with Codex. For example, PFA does not cover rules for the various tests for water as required under Codex.

According to some small exporters, HACCP has not been followed in the pulp industry! There is some general awareness about HACCP, but they think it has not yet been passed as a law, and they do not have to worry about it yet, especially because there is no consumer insistence in India for such standards. While they admit that HACCP will certainly increase market accessibility, they believe they will face several problems in adopting it. Some of the problems pointed out were:

(a) Since orchards land holdings are small and contractors procure the raw material, it will be impossible to keep records at the field level as required for HACCP. The general age of orchards ranges between 3 - 100 years, so it will be difficult to establish control;

(b) Since this industry is seasonal (3 months) it is not feasible to adopt these standards, and to retrain staff, as the units cannot keep permanent staff. Training new staff every year is also too costly;

(c) It will be more viable for large plants or industrial houses, which deal in multiple products, work throughout the year and have their own orchards. But most of the units are small in this sector and HACCP will not suit them;

(d) As far as the financial aspect of HACCP compliance is concerned, units that are setting up now will not have any serious problems. It does not cost much for new units to be compliant, but the old units have to revamp their infrastructure. It is a costly affair; according to rough estimates, HACCP compliance will raise costs by 40 per cent;

(e) Financial institutions do not fund HACCP activities;

(f) Main markets for Mango Pulp are Gulf countries, and they are only interested in cheaper prices not HACCP; and,

(g) It is quite costly to get ISO certification. The cost may range between 1.5-2.5 lakhs for ISO audit. The surveillance audit is every six-months and it costs Rs. 10,000 per manday.

Apart from HACCP, pesticide residue affects Mango Pulp industry in the same way that it affects peanuts. Other quality issues are, that Indian pulp is brown in colour, is supplied in punctured bags, drums in which it is exported are of poor quality, experience feathering (peeling of the coat), rusting, metallic taste (tin taste) and damage to seam of the tin or drum. These are essentially packaging issues. The packaging problem identified above arises from the low quality of packaging material available in the domestic market. While imported tin is of good quality, it is more costly.

According to exporters, they either do not have the technology, or the technology is too costly and they do not have the economies of scale to meet these costs. They feel that packaging should not be considered a health hazard. Testing is a major problem for these units. There are a number of institutions but these are spread all over the country and their charges are quite high. The Central Food Technological Research Institute (CFTRI) charges Rs. 3000/- per test and Societe Generale De Surveillance (SGS) charges 0. 27 per cent of f. o. b. value of the consignment. Most laboratories in India do not have the sophisticated equipment required to carry out the increasingly more complicated tests necessary to comply with HACCP.

For example, foreign health authorities are moving from testing for parts per million (ppm) to parts per billion (ppb). Indian laboratories are not equipped to do these tests. There are differences between the test results of India and those of Europe, allegedly due to the methods of testing, and not due to different test objectives. In Europe only 'natural food' imports are encouraged, i. e. no sugar should be added, though sugar is also a natural product. If sugar is added, import duty increases by 13 per cent (i. e., from 6. 5 per cent without sugar to 19. 5 per cent with sugar). However, they do add sugar themselves in Europe, using their beet sugar surpluses (generated by farm subsidies in Europe).

All the former colonies of France, Portugal and Spain do not pay duty on food items into EU, but all the former colonies of UK have to pay duty. Successful exporters feel that the quality of Indian food has to be monitored for exports, and APEDA should introduce licensing. But it is very difficult to monitor implementation of norms if everybody is allowed to export and they argue that small scale units should not be encouraged to export because of their fly-by-night operations. The FPO has issued 4700 licenses for food processing units, of which 21 are large units, 156 medium and the rest are small scale; 90 per cent of these units produce mango pulp.

Exporters often have good long term relations developed and established over many years with certain buyers. In the event of any trouble with the authorities on account of quality or any other reason, most such buyers are helpful in sorting out the problems at their end, due to their stake in the clearance of consignments. The quality issue becomes a major hurdle when the buyers have excess stock or when the international market prices of the goods have fallen below the agreed/contracted prices. In such cases, sometimes the exporters have to accept price discounts, especially when goods are perishable. There are several issues with government agencies.

The Ministry of Commerce takes an interest in the exporters operations, as they are responsible for trade promotion. But if the problems faced by the exporters are quality or health related, it is the Health Ministry that should be involved. Even in business negotiations, foreigners want an assurance from the Health Ministry, which is not easy to obtain. There is clearly a need to create better policy coherence here. The Health Ministry is responsible for the development of Codex standards.

The exporters feel that the Ministry could benefit from consultations with producers when attending Codex meetings and formulating domestic standards. Food laws lay an emphasis on economic offences and not on safety; hence the basic thrust of food laws is misplaced as far as export promotion is concerned. On a positive note, the study clearly indicates the positive impact of pro-active measures by an apex export agency like APEDA in enhancing quality export earnings.

Mushrooms

Trade in Mushrooms has gained importance in recent years for two main reasons, namely; (i) the global shift towards vegetarian food, and (ii) Mushrooms are a rich source of protein and other nutrients that lends itself to industrial applications. A third dimension could be added in the Indian context, given its rich bio-diversity, and this may be more generally applicable for developing countries that share this attribute. Mushroom production has strong backward linkages in terms of employment generation, without competing for scarce arable land resource. In fact, mushroom cultivation offers an opportunity to add value add to some coarse and inedible biological resources, with a little help from science.

India has been exporting both fresh and dried Mushrooms on a small scale for several years, but recently exports have expanded. Fresh Mushroom exports increased from 790 kgs. In 1990'91 to 11. 8 million kg. in 2001/02. During the same period, dried Mushrooms exports reached 0. 242 million kg. During 1990/91-2001/02, fresh Mushrooms recorded a growth rate - in value terms - of 3. 43 per cent, while dried Mushrooms grew at 3. 11 per cent. USA was the major (97. 3%) importer of fresh Mushrooms in 2001/02. The main destinations for dried Mushrooms during 2001/02 were, USA (54%), Germany (19. 1%), Switzerland (16. 7%) and France (4. 6%), with Canada, Denmark, Swaziland and Belgium accounting for the remainder.

The major markets for dried/processed Mushrooms from India prior to 1997/98 were USA, Israel, Denmark and Canada. At present, most of the small growing units are experiencing severe constraints to achieving higher productivity levels. Quotas on mushrooms in the EU are an uncorrected legacy of the past, when the agriculture sector was not covered by GATT disciplines. The tariffication process built into the Agreement on Agriculture of the WTO is yet to result in quota free access of Mushrooms to EU.

Multilateral efforts are required to expedite this process. Meanwhile, India needs to make representations to the EU to obtain an exclusive quota. The quotas allocated to Poland and other countries have been regularly left under-utilised to the extent of 21,000 to 22,600 tonnes in the last couple of years. India should present its case to fill-up such under utilised quotas through a separate allocation. Mushroom growing in India started with the use of primitive technology for compost making/ crop raising in the late 1960s/early 1970s, which resulted in low yields per unit weight of compost. The compost was prepared from cereal straws and animal waste by a long outdoor fermentation process in a single phase without use of steam pasteurisation.

An average Mushroom yield 6-8kg/100 kg compost was harvested in 6-8 weeks of cropping and the crop was raised in make-shift cropping rooms. This was followed by the second phase of activity, when a modern mother composting unit was established in Solan with FAO assistance. Compost was prepared here by an improved method in 2 phases, and a rich substrate was prepared from cereal straws/poultry manure. This richer compost doubled the productivity of mushrooms per unit weight of compost, which is considered a big leap for the growing mushroom industry in our country.

With the increased exposure of scientists/workers to modern growing methods, and with more and more people taking to this profession, the Mushroom industry started to mature. The establishment of the National Research Centre for Mushrooms (NCRM) by the Indian Council of Agricultural Research (ICAR) at Solan, H. P. in 1983, further provided fillip and encouragement to the industry. Government support for R&D in this sector supplemented concerted efforts for popularising improved methods of cultivation, screening of improved strains for use by seasonal growers, and addition of more varieties of Mushrooms to the list of cultivated Mushrooms in India.

Information on improved technology was still not available to the average common grower or an upcoming entrepreneur in India. Thus, though the grower produced the compost, there was a bottleneck on accessing information on raising a healthy crop of mushrooms. The grower would collect spawn from some source but not know about the growing parameters, nor was the modern cropping room available to him. So, the majority of the growers continued to use this method of growing in improvised cropping rooms till the late 1980s, resulting in poor yields.

In the late 1980s and early 1990s, modern cultivation units were established with help from various companies in Europe, which were more interested in selling their machinery and in the establishment of the mushroom farms at their asking rate. This did help in the building of modern mushroom units, but the big question was who would manage and run them. That is where the Indian industry took a beating, and unit after unit failed to produce mushrooms on a profitable scale. Producers took some time to tune the production parameters until most units could obtain economic yields. But, by the time this was achieved, the international market came crashing down. This situation is still continuing.

The available R&D support caters more to the information and training needs of the small/marginal Mushroom growers. The average yield per unit weight of compost has been increased to 16-20 kg/100 kg compost in 6 weeks of cropping. But to become globally competitive, the aim is to achieve yield increases combined with reductions in cultivation costs.

The foreign machinery sellers painted a rosy picture of the market. They applied the technology and machines, used in labour starved countries of Europe, on an 'as-is-where-is' basis, and the results were not very encouraging. The machinery sellers from industrialised nations, besides selling machinery, also offered technical know-how for the cultivation of the white button Mushroom in cropping rooms with a computer-controlled environment. They failed to understand that the conditions in temperate Europe and in tropical India are fundamentally different. While the temperature needs to be raised under European growing conditions, the Indian requirement is the opposite.

The task becomes even more complicated when not only temperature, but also other parameters, like R&D, air speed, heat removal, CO/O content of the cropping room need to be manipulated. All such parameters must be maintained at specified levels during various stages of crop raising; as change in a given parameter affects the others. Further, most of the raw materials used in Europe are not available in India. For instance, peat for casing is not available in India and alternative materials have to be used in its place. The European grower is used to heavily watering the peat casing, which is not relevant for Indian conditions.

Post-harvest handling of fresh Mushroom in temperate areas is easier as compared to warmer climates where there is a premium on quick post-harvest handling. The foreign machinery sellers offered buy-back arrangements to most of the projects (exportoriented units), but in-fact the end result was a fiasco. This was a false guarantee given to naive entrepreneurs to induce them to participate in the project, a fact which the entrepreneurs realised too late. It became fashionable for financiers in

India to ask for a buy-back guarantee from a foreign buyer, which in the opinion of the buyers was nothing but a ploy to safeguard the interests of the financiers.

What would have been genuinely useful would have been measures to ensure that every project entrepreneur obtained a clear picture of the national and international marketing system and the market place. Every project entrepreneur should undertake a market survey on a realistic basis, and provide his assessment and projections of the market. Of course export markets are not the only outlets for Mushrooms; India itself will be a big market for Mushroom growers in the future, especially for fresh marketing of the produce.

The raw materials available for mushroom cultivation vary in different parts of India. In most of the northern and central parts of India, wheat straw is widely available but at high prices. In the eastern and southern parts, paddy straw is available in abundance, and at lower prices. Poultry manure is available everywhere at a very low price. Sugarcane bagasse is available in those areas where sugarcane is widely grown (western parts of India, central India, and some other places). The requirements and techniques for composting differ with different base materials (wheat straw, paddy straw/ or sugarcane bagasse) and information available to growers on the use of paddy straw/ or sugarcane bagasse as a base material is limited.

Optimal utilisation of the above materials for composting to obtain economic returns requires specialised skills from the manager/entrepreneur. Casing is the second important input in button Mushroom growing. Because not many casing materials are available for commercial growing in India, the choice has narrowed down to Farm Yard Manure (FYM)/spent compost/ composted coir pith, etc. These materials require to be processed (water leached/steam pasteurised) before use, unlike the peat sold in Europe that is harvested from underground bogs deep down. This material is devoid of harmful micro-organisms and is used after adjustment of pH with lime. Use of the above agrowastes as casing in button Mushroom cultivation, in place of peat, again requires experience on the part of the grower for their optimal utilisation. Though NRCM has generated information on composting/casing material usage, and Mushroom productivity from these materials on a commercial scale, the price factor is indeed critical.

There is no organised assistance available for marketing of this produce in India. Every export oriented unit has its own individual arrangement for marketing, and the Mushrooms are preserved in brine and canned in large containers of 3-5 litres (or bigger) capacity for export. Government support for Mushroom marketing is not available in India, nor is any special/preferential quota available in the European Union (as is available to certain other nations). Direct export to USA/Germany under some sort of arrangement is one alternative that could be considered. For this, growers will have to form a marketing co-operative.

There are no processing plants devoted to providing support for this industry in India, except for some limited support by NAFED to seasonal growers in the north-western plains of India. Such support is available to Mushroom growers in China, where they are able to can the produce on a large scale at rail-accessible points for export. Finance at lower interest rate, and inputs for infrastructure at fair prices, should also help this industry to keep down the cost of cultivation. Lower production cost, together with greater productivity per unit weight of compost, can help the industry to become globally competitive.

A long-term strategy has to be developed to help the industry. This could include manpower training, development of high yielding strains, a better pest management programme, and efficient post harvest handling/processing systems for value addition. Clearly, in the case of Mushrooms, India currently faces unfriendly tariff structures and quota issues in world markets on the one hand, and capacity and technology issues on the other hand. These, rather than environmental issues, are the major constraints to the development of the industry.

8

Food Processing Policy 2005

I. Preamble

1.1 The farming sector has come a long way since independence. From a country beset with shortages India has attained self sufficiency in food with sectoral surpluses emerging in production. With the growth of the economy a shift is also being seen in the food basket from consumption of cereals to a more varied and nutritious diet of fruit & vegetables, milk, fish, meat & poultry products. These changes have brought in their wake the imperatives of an emerging industry of Food Processing Industries.

1.2 India with arable land of 184 million hectares, produces annually 91 million tonnes of milk (highest in the world), 150 million of fruits & vegetables (second largest), 483 million livestock (largest), 210 million tonnes foodgrain (third largest), 6.2 million tonnes fish (7th largest) and is the fifth largest in egg production. Though yields and per capita consumption patterns are expected to improve, yet moving from subsistence levels of production to a marketable stage has brought in pockets of surpluses which has led to wastages of produce. Surpluses have led to localised marketing difficulties and at times a fall in prices with consequent impacts on farmer incomes. Inspite of surpluses and wastages our levels of processing has been low.

1.3 Conservative estimates put processing levels in the Fruits & Vegetables Sector at 2%, Meat & Poultry at 2%, Milk by way of modern dairies at 14%, Fish at 4%. Bulk meat de-boning is to the tune of 21%. Processing in the areas of foodgrains, oilseed, tea, coffee, etc. where raw products have perforce to be processed have had a tradition of being processed is high. Even in these products the demands of modern food processing sector are still emerging. Compared to the developed countries of the world we are far behind, unfortunately we lag behind even compared to countries like Philippines and China where value addition is 45% and 23% respectively as compared to our rate of 7% in food products.

1.4 The food processing sector though in the nascent stage constitutes 14% of manufacturing GDP and amounting to products value of Rs.2,80,000 crores. It employs 130 lakh persons and has the propensity to promote 2.4 times the indirect employment of its direct employment creation. Employment over value added is larger as compared to the manufacturing sector as a whole. Over the last decade or so food processing has grown at a rate of 7.1% per annum. The higher rate of growth as compared to the Agriculture growth rate is indicative of its low base, the increased availability of surpluses, changing life styles, tastes and higher disposable income with consumers. It is expected that the annual growth rate in the next five years shall be to the tune of 7.3% per annum.

1.5 The area of responsibility of the Ministry of Food Processing Industries encompasses grain milling, milk, meat, poultry, fish confectionary, bakery, convenience foods, and fruit & vegetable processing sectors, aerated drinks, beer & non-molasses based alcoholic drink. These industries as a group have a turnover of 1,50,000 crores and have shown a higher rate of growth. They are expected to maintain this trend and grow at the rate of 7.3% per annum in the next five years. This policy is with regard to areas of responsibility of the Ministry.

1.6 With the advent of the WTO regime and the possibility of lowering of subsidies in the developed countries, our competitive advantages in food production and processing industry shall come to the fore. However, due to a variety of factors though world trade has moved more and more towards processed commodities, our export has performed unevenly with regard to such value added products.

II. Objectives & Goals

2.1 The stepped up growth of food processing should lead to the fulfilment of:

a) Higher farmer incomes and the providing of a larger as well stable market for farmer produce.

b) Reduction in wastage of farm produce.

c) Provision of nutritious, safe food and wider consumer choice.

d) Enhancing employment opportunities by way of direct and indirect employment.

2.2 With a view to achieve these objectives we should aim for doubling the food processing sector of in the next five years i.e. aim for a compound annual growth of nearly double of what was achieved in the last five years. The stepped up growth shall have major beneficial impacts in terms of the objectives for the food processing policy. Industry alongwith government and concerned institutions shall need to act in concert to achieve these goals.

III. Constraints & Strategies - Across Sectors

Affordability & Costs of Processed Food

3.1 In the developed countries processed food and fresh food prices are usually within a reasonable comparative range. At times processed foods are even cheaper than fresh food. In India, however due to a variety of factors processed foods prices are substantially higher than fresh food. Given the objectives which are achieved by processing there is an urgent need to take measures to reduce costs, and make processed food affordable.

3.2 Of late there has been a draw down of taxation for the perishable sector, alongwith Income Tax waivers for fruits & vegetables. Reduction/waive in excise has also been forthcoming. The whole perishable sector should be treated in a similar fashion, the losses to the exchequer shall be notional. Cost of packaging is the other major constraint of this sector. Cost of packaging ranges anywhere from 10 to 64% of production costs. The high level of intermediation in the Indian food chain also pushes up costs.

Efforts should be made to assist towards consolidation of the chain and the reduction of transaction costs in this regard, lack of infrastructure, post harvest losses, multiplicity of regulatory authorities also add to costs of the processed foods. Some of the dimensions outlined above have a linkages, constraints, and remedies falling in other functional areas and shall be dealt with under respective groupings.

Processors, Farmers & Agri - varieties

3.5 Linked with the issue of fostering relationships between processors & farmers is the possibilities of varietal relevance and productivity increases. Agriculture often suffers from varietal constraints. Oranges in Nagpur and

Mangoes in Malihabad (UP) are instances where though there are gluts at harvest time processing has not taken off due to want of processable varieties.

3.6 The amendment/modification of the Agriculture Produce Marketing Commodities Act would facilitate the reduction in costs, as well as, allow the fostering of a beneficial relationship between the relevant players. Contract farming is an intervention which appears to have helped the growth of the processing industry as well farmer incomes. Care shall however need to be exercised to strike at the correct balance between the farmer and processors rights and obligations.

3.7 The traditional extension machinery has at times being unable to deliver on varied demands and requirements of farmers. Some states have fostered public-private relationships in this regard. Others have partnered with company led initiatives like e-choupal and the encouragement of contract farming. It shall be the endeavour of the policy to bring forth the best practises being adopted before the States for them to exercise appropriate choices in the field of extension and linkage with processors and research institutions.

Supply Chain & Post Harvest Issues

3.8 A characteristic of the Indian food chain is that it is a very long and fragmented chain, which adds to cost, and at times a disconnect between producers and consumers. The chain today leads up to aggregation at the mandi level and then disaggregation to the consumer. Other issues specially for fresh produce would be better addressed if organisations large food retailers or markets come into being which procure directly from farmers or at best villages and make it directly available to consumers.

3.9 Besides the observed fragmentation of the supply chain we need to improve on post harvest handling, cleaning, grading, packaging and storage. Storage protocols need to be developed for the varied agri produce available. Presently work has been concentrated on first a few crops like wheat, rice, grapes, mangoes, citrus and capsicum.

Infrastructure Support

3.10 Infrastructure for this sector is lacking. Deficiencies exist for grading and packing besides pre-cooling at farm site which could feed into a formalised cold chain. The cold storage capacity today caters to less than 10% of the produce, this too is of rudimentary nature with over 80% designed only to handle potatoes. A paucity of chilling infrastructure for milk and a lack of modern abattoirs for the meat processing sector is seen. Fish processing more specifically for exports requires a major step up in infrastructure availability. Physical marketing and warehousing infrastructure also needs to be upgraded.

3.11 India has limited Controlled Atmosphere Storage facilities, technologies, protocols and machinery is chiefly imported. While no attempt is suggested to reinvent the wheel, we should develop expertise to assess and modify and if necessary upgrade manufacture capabilities according to our specific requirements.

3.12 These areas need support and new paradigms of private-public partnership. Government subsidy schemes, partnership with banks, creation of models for JVs, SPVs for implementation as well as dedicated infrastructure financing institutions are required to give this area the requisite push.

3.13 Besides the hard production infrastructure gaps mentioned in the para above gaps exist in the Lab. Research and training infrastructure which have already been mentioned in respective functional areas.

Financing Issues

3.14 Within the priority sector lending target of 40% the banks are mandated to lend 18% to direct and indirect Agricultural activities. Food Processing finds place within a further subcategory of indirect lending of 4.5%. This sub-category is usually oversubscribed by banks as even agriculture inputs find are a part of this sub-category. Directed lending has its own limitations in expanding credit to this sector. Still removing conditionalties of caps on financing, allowing advances for crop purchases to be considered as part of this lending shall give a signal and assist this sector in improving its financial availability and muscle.

3.15 Agricultural produce is seasonal in nature and thus subject to gluts more specifically in the perishable sector. A difficulty in this endeavour are the finance needs of the farmers, processors. Warehousing receipts should therefore be made negotiable instruments. A system of accreditation of warehouses would need to be put in place so that receipts gain value and are tradeable. Issues coming in the way of this mechanism should be resolved.

3.16 Prices of agricultural produce often fluctuate leading to confusing market signals to farmers and upsetting cost estimates of processors. To make commodity markets more stable the future markets need to be deepened. Banks, Mutual Funds and FIIs should be allowed futures participation and trading Commodity based derivative funds and trades should also be allowed.

3.17 Liberal funding for infrastructure development under RIDF would also provide the much need funds to give a leg up to this sector. States could then be well advised to go into Joint Ventures to build and operate such infrastructure.

3.18 Commercial banks have substantial flexibility in lending. However practises on risk assessment vary a great deal which leads to an uneven performance. Best practices of banks should be studied and disseminated to enhance credit flow to the sector.

3.19 New and innovative methods should be identified to enable structured term lending for infrastructure development.

Research Development & HRD

3.20 Institutions need to engage both fundamental and applied research and keep abreast of global demands. Based on preferences of Indian consumers they need to address parameters of physical, chemical, taste, flavour & spoilage. Research needs to be further charaterised by offering scale-able models. Institutions should also be charged with patenting indigenous technology as well as technology they generate. An effort therefore, to begin with, needs to be undertaken to network the present institutions and establish a database on the present availability of technologies, patents and capabilities which shall serve as the base for further work in this field.

3.21 India has a very wide range of eating habits and food products. Such products and processes need to be documented where necessary improved upon and patented. R&D on ethnic foods shall not only address regional needs but open up a market amongst the large Indian diaspora.

3.22 Looking at industry and users' demands it would appear that steps need to be taken to further nurture high quality institutions in this field. An apex institutions could be established which would inter alia provide for world class graduates and research programmes; carry out cutting edge research; provide for a framework for a constantly updated curricula to sister institutions, assist and provide a linkage to regional institutions in their activities.

3.23 A special area of concern is the packaging sector. High quality packaging today requires costly machinery as well as costly materials. Research on packaging needs to be stepped up to arrive at cost-effective solutions.

Food Safety & Hygiene

3.24 The Indian emphasis on fresh food consumption is also linked to the quality perception of processed foods. Processed foods by definition imply a certain shelf life before consumption. Unless we are able to assure the consumer that this storage has not led to a deterioration in food quality the consumer would shun processed foods. Consumers therefore look to safety in certifications and brands, where they do not have an intimate knowledge of the product or process. Certifying authorities therefore need to be vigilant in enforcing standards. Quality enforcement staff should also be fully trained on HACCP, GMP and GHP requirements and measures. A well equipped and certified lab. infrastructure needs to be put in place commensurate with a hierarchy of needs at the regional state and divisional/district level as the need be.

3.25 Institutions need to develop expertise on standards and CODEX issues. Efforts should be made to harmonise national standards with CODEX standards so that our populace is assured of good quality product. CODEX should form the starting point for making risk assessment and analysis to help arrive at science based standards, relevant to Indian conditions.

Exports

3.26 Processed food exports account for just 1% of trade in this sector have grown at an average rate of 7% per annum for the last five years accounting for about 70% of total agricultural exports. Issues of quality standardisation and certification are accentuated in this field. The rate of growth needs to be doubled over the next five years. Export related infrastructure agri produce specially at sea and airports needs to be strengthened. Brand building, market intelligence and a continuous watch on areas of emerging competitiveness needs to be maintained to enhance exports.

Legal Framework

3.27 Issues of quality and standards have already been touched upon. The legal framework governing the sector is in need of change. Nine ministries and over thirteen central orders lay down the law, rules & regulations, just at the Union level. States have their various control orders as well. The organisations responsible for these regulations are often poorly staffed, trained and have diverse objectives and orientation. Consequently industry is faced with an inspector raj. Recognising this issue the Union Government has set up a Group of Ministers to consider an Integrated Food Law. This effort would go a long way to bring in a law which balances the imperatives of food safety, consumer choice, industry representation and compliance.

3.28 Processing industry is by nature seasonal. Assets remain idle over a prolonged period of time. Karnataka in recognition of this character has given it a status which exempts it to an extent from labour, and other regulations. Efforts therefore need to be make to see how relief can be extended given its seasonal nature.

3.29 Scale of industry is determined by economic imperatives of markets and even without reservation of the overwhelming number of items over 80% of this industry is in the small scale sector. There has been no evidence of the collapse of industry or unemployment in the items dereserved over the last decade or so. While there has been progressive dereservation of processed foods six categories still remain reserved for production in the small scale sector. A measured dereservation of the residual categories needs to be followed.

3.30 Contract farming is one of the measures which would facilitate processing. Some states have already provided for this as an alternative method of procurement. A relook of the APMC Acts in the states is urgently required to allow choices and alternative modes of sourcing to entrepreneurs.

Support Structure

3.31 Though the share of Agriculture GDP has reduced from 62% to 22% from 1954 to 2004 yet over 66% of our population draw sustenance from this sector and, of course the sector is of concern to consumers encompassing the whole population. The large canvas of the sector and at times crisis in particular subsectors has spawned a large number of interventions.

Alongwith the regulations Government over a period of time has created a very large number of promotional organisations. Separate commodity boards of Tea, Coffee, Cashew, Coir, etc. besides, APEDA, MPEDA, EIC, NHB, CFTRI, DFRL, IIP, D.M.I., Ministry of Agriculture and Ministry of Food Processing Industries to name a few exist. Just as it is time to consolidate the legal framework it is also time to consolidate the promotional framework to obtain optimum results both at the Union as well as the State level.

IV. Constraints & Strategies - Sectoral Issue

4.1 While most of the issues are cross sectoral in nature some sectors do have specific issues to be addressed. In the fruit and vegetable sector to enhance exports a programme to lift quarantine measures inposed by certain countries should be pursued. For certain fruits or their pre-shipment treatment needs to be taken in hand. Issues of organic produce and awareness of systems in place needs to be disseminated.

4.2 In the dairy sector R&D an indigenous products needs greater effort. The milk procurement and testing has to devise means to go beyond fat content testing at collection points to issues of SNF, bacterial count, somatic cell count and residues.

4.3 In the meat & poultry sector one would need to concentrate on disease free zones, partnership structures for abattoirs and the removal of deficiencies in the feed supply chain.

4.4 The fishery sector and more specifically the marine sector has vast possibilities. Processing infrastructure for sea catch needs to be strengthened.

4.5 In the foodgrain sector we would need to look at storage and bulk handling issues for the domestic and export sector and look at means to export value added products instead of just grains.

V. Special Category States

5.1 Special category states are by and large less developed in economic terms. Often characterised by a weak industrial base and chiefly agrarian, food processing offers possibilities of development in their case. Strategies should be tailored to their specific needs, so as to reap benefits through food processing initiatives.

VI. Policy Review

6.1 The pace of scientific, social and economic change has accelerated. Policies need to address these changes to guard against threats as well as take advantage of opportunities. To ensure that policy remains in tune with its times, a review every five years in consultation with concerned stakeholders should be undertaken.

9

Safety Evaluation of foods

Progress in biotechnology will play an increasingly important role in food supply. Recent years have seen tremendous advances in food biotechnology, including improvements in industrial process technology and control systems, improvements in farming systems for growing and harvesting food, genetic improvements to organisms used in the food supply, and improvements in techniques to monitor food safety and nutritional quality.

Traditional Food Biotechnology

Examples of traditional food biotechnology include the use of yeasts in the brewing and baking industries, and the use of bacteria and moulds and their components in the dairy industry for making cheese and yoghurt. Moulds and bacteria are also used for the fermentation of plants or plant products (for example, miso). Purified enzymes from micro-organismsare used extensively in making products such as high-fructose corn syrup and certain types of hydrolysed or predigested protein products.

In many such products, the micro-organisms function in the production process and the food product does not contain viable cells. In others, such as yoghurt, microbial cultures remain viable and are consumed. Such traditional applications have a long history of safe use, and many have formally been affirmed as safe by various national and international food safety evaluations. Key considerations have included non-pathogenicity and non-toxicity of the organism and its products. Modem biotechnologies are being used increasingly to improve food micro-organisms for the enhanced production of essential components or products, as well as the improvement of nutritional value, flavour, texture, and the shelf life of fermented foods.

Plants

Plants are consumed directly as whole food, or are processed into many types of foods. Many plants have a long history of use as foods. Undoubtedly, the plants selected were the ones that appeared healthy, grew vigorously, and gave higher yields. Edible portions had desirable taste, smell and appearance. Selection might have included an evaluation of safety, although it was not formally recognised. In any case, there is little historical record or documentation of the process by which the safety of food plants was maintained, or of involvement of national food authorities.

Now that new biotechnology has vastly increased the variety of new traits that can be introduced into plants, the impact of plant biotechnology on food safety is receiving attention. Early farmers selected and preserved plant variants that had desirable food or agronomic attributes, such as larger fruit or uniform dormancy and maturation times for seeds. Such properties are deleterious to wild plants and so would not have been developed without the efforts of early "breeders". Practices of early farmers led eventually to the development of desirable clones, land races, and varieties of major food crops, with predictable reproducible agronomic characteristics yielding foods with uniform properties.

As the agronomic properties of individual crops were made more uniform, production methods could be designed to obtain optimum yields. With the relatively recent advent of directed plant breeding for improvement of agricultural crops, the objectives of plant breeders became:

i) to increase yield;

ii) to improve quality; and

iii) to reduce production costs by, for example, identifying traits which could increase resistance to pests and diseases.

Although it may not have been a major objective, plant breeders have been effective in conserving the nutritional quality of plants developed for food. Routinely, they have selected plants with desirable qualities and rejected undesirable plants by destroying them in the breeding plots. Preferences of the humans consuming the crop have contributed to the food characteristics of plant varieties ultimately developed. For example, varieties of potatoes and beans are quite different in different areas of South America, where their selection has been influenced by the taste preferences of native peoples.

As another example, the milling and baking quality of flour is checked during the variety development process since wheat is often developed for particular baking products. In the case of certain crops, breeders have deliberately attempted to improve nutritional value. Often, as for example in the case of high-lysine corn or high-vitamin C tomato, other factors have prevented these varieties from becoming widely accepted. The best-tasting, most nutritious variety will not succeed as a commercial crop unless it also gives high yield. Difficulty in processing, susceptibility to pests or diseases, an undesirable flavour or colour, or simply difficulty in getting the plants to market will also limit the adoption of a new variety. Public acceptance of a high-nutrient variety is not based on nutrient content alone.

Carrot and sweet potato varieties that have a bright orange colour are more acceptable to humans than those that do not. They also have a higher content of the pigment that supplies vitamin A precursor in the human diet. The ascorbic acid (vitamin C) content of tomatoes has been extensively examined, and varieties with higher content developed. However, since the fruit of these tomatoes is more yellow-orange than red, they have not been as acceptable to consumers. The nutritional value of fruit or vegetable crops can be quite variable, and may be difficult to assess definitively.

The composition of plant foods, particularly fruits and vegetables, is transient because the edible portion undergoes rapid biochemical changes during the ripening process. For example, in red tomatoes the content of ascorbic acid is low in green fruit, increases rapidly as the fruit ripens, and then drops off with time. Ascorbic acid content also varies in ripe tomatoes with their position on the vine, since higher light intensity increases the amount. Moreover, field-grown tomato plants produce fruit of higher vitamin C content than those grown in glasshouses. In view of these considerations, the significance of a genetically induced change in the level of a nutrient, such as ascorbic acid in modified tomatoes, would be difficult to assess.

The significance of a genetically induced change in the level of a particular nutrient would also depend on the position of the food in the total diet. Many plants are known to produce compounds toxic to other species. Acutely toxic poisonous plants, such as some fungi and ornamental plants, are not consumed. A number of plants consumed by humans are acutely toxic in the raw state, but are accepted as food because processing methods alter or eliminate their toxicity. For example, the cassava root is quite toxic, but proper processing converts it into a nutritious and widely consumed food. Soybeans and lima beans, among other crops, also require proper processing. Thus the mere presence of a toxicant in a plant variety does not necessarily eliminate its use.

In other plants that contain toxicants affecting humans, such as potato and tomato, plant breeders have succeeded in reducing the level of these toxicants in food varieties. Over time, there have been few reported examples of plant breeding inadvertently leading to increases in toxicants. Varieties with an increased toxicant level have been quickly removed from agricultural use. In some countries new varieties have been monitored for levels of a particular toxicant, but systematic food safety assessment has not generally been conducted.

The impact of plant biotechnology on food safety is now receiving wider attention. At the same time, there is increasing general recognition of the relevance of plant breeding's historical record. Toxicant levels might become important, particularly when traits are introduced for resistance to pests and diseases, simply because a compound inducing resistance to another organism might possibly affect humans.

The molecular basis for the resistance mechanisms is just beginning to be understood by plant scientists, and may be a target for biotechnology approaches to enhance resistance. Some mechanisms appear to be quite general, while others have adverse effects on a specific pest or pathogen. Knowledge of the mechanisms should, in the future, provide a valuable tool for the plant breeder and should facilitate evaluations of safety.

Animals

The development of new strains of domestic mammals and birds for food has had a long history, and extensive procedures are in place to improve yield and assure the health of these animals. In general, foods from new strains of mammals and birds that appear to be in good health have proven to be as safe as the animal breeds from which they were derived. No endogenously produced toxicants are known to come from such domestic animals.

In recent years, breeding technologies have been developed that permit increased numbers of desirable individuals through techniques such as embryo splitting. In addition, improved knowledge of the genetic control of hormonal levels has permitted the alteration of carcass quality, for example of fat to lean ratios, which has resulted in consumer-desired lean meats. Increased hormone levels have also enhanced the rate of growth, as well as milk production. There is no evidence of adverse effects to humans from the use of such technologies.

Concepts and Principles of Food Safety

The consideration of the safety of foods and food components derived from biotechnology involves several continua: from older to newer biotechnology; from traditional techniques to the latest techniques based on molecular and cellular biology; from simple to complex products; from a well-known history of exposure and safety of use to areas of less knowledge of the trait in different organisms; from whole organisms to specific chemical compounds or substances; and from simple to complex assessment approaches.

For a rational and practical approach to ensuring safe use, these continua can be separated into manageable pieces, facilitating the description of the concepts or principles of safety. Accordingly, scientific principles and procedures should be applied in a flexible fashion, taking into account the knowledge of: the characteristics of the newly introduced trait(s); potential dietary exposure; the preparation and processing of the foods or food components; nutritional considerations; and toxicological aspects.

The safety of food for human consumption is based on the concept that there should be a reasonable certainty that no harm will result from intended uses under the anticipated conditions of consumption. Historically, foods prepared and used in traditional ways have been considered to be safe on the basis of long-term experience, even though they may have contained natural toxicants or anti-nutritional substances. In principle, food has been presumed to be safe unless a significant hazard was identified. Modern biotechnology broadens the scope of the genetic changes that can be made in food organisms, and broadens the scope of possible sources of foods. This does not inherently lead to foods that are less safe than those developed by conventional techniques.

Therefore, evaluation of foods and food components obtained from organisms developed by the application of the newer techniques does not necessitate a fundamental change in established principles, nor does it require a different standard of safety. Moreover, the precision inherent in the use of certain molecular techniques for developing organisms for use as food should enable direct and focused assessment of safety where such assessment is desired. Knowledge obtained using these methods might also be used to approach safety assessment of new foods or food components from organisms developed by traditional methods.

For foods and food components from organisms developed by the application of modern biotechnology, the most practical approach to the determination of safety is to consider whether they are substantially equivalent to analogous conventional food product(s), if such exist. Account should be taken of the processing that the food may undergo, as well as the intended use and the exposure. Exposure includes such parameters as the amount of food or food component(s) in the diet, the pattern of dietary consumption, and the characteristics of the consuming population(s). This approach provides a basis for an evaluation of food safety and nutritional quality.

The concept of substantial equivalence embodies the idea that existing organisms used as food, or as a source of food, can be used as the basis for comparison when assessing the safety of human consumption of a food or food component that has been modified or is new. If one considers a modified traditional food about which there is extensive knowledge on the range of possible toxicants, critical nutrients or other relevant characteristics, the new product can be compared with the old in simple ways. These ways can include, inter alia, appropriate traditionally performed analytical measurements (for example, alkaloid levels in potatoes, cucurbatin in vegetable squash cultivars, and psoralens in celery) or crop-specific markers, for comparative purposes.

The situation becomes more complex as the origins/composition/exposure experience decreases, or if the new products lack similarity to old established products or, in fact, have no conventional counterpart. A demonstration of substantial equivalence takes into consideration a number of factors, such as:

— knowledge of the composition and characteristics of the traditional or parental product or organism;

— knowledge of the characteristics of the new component(s) or trait(s) derived, as appropriate, from information concerning: the component(s) or trait(s) as expressed in the precursor(s) or parental organism(s); transformation techniques (as related to understanding the characteristics of the product) including the vector(s) and any marker genes used; possible secondary effects of the modification; and the characterisation of the components or trait(s) as expressed in the new organism; and

— knowledge of the new product/organism with the new components or trait(s), including the characteristics and composition [i.e. the amount of the components or the range(s) of expression(s) of the new trait(s)] as compare with the conventional counterpart(s) (i.e. the existing food or food component).

Based on a consideration of the factors in the paragraph above, knowledge that a new food or food component(s) was derived from organism(s) whose newly introduced traits have been well-characterised, together with a conclusion that there is reasonable certainty of no harm as compared with its conventional or traditional counterpart, means that a new food or food component(s) can be considered substantially equivalent. Set out below are the principles for the application of substantial equivalence to the assessment of foods from organisms developed by the application of biotechnology:

— If the new or modified food or food component is determined to be substantially equivalent to an existing food, then further safety or nutritional concerns are expected to be insignificant;

— Such foods, once substantial equivalence has been established, are treated in the same manner as their analogous conventional counterparts;

— Where new foods or classes of new foods or food components are less well-known, the concept of substantial equivalence is more difficult to apply; such new foods or food components are evaluated taking into account the experience gained in the evaluations of similar materials (for example, whole foods or food components such as proteins, fats or carbohydrates);

— Where a product is determined not to be substantially equivalent, the identified differences should be the focus of further evaluations;

— Where there is no basis for comparison of a new food or food component, that is, where no counterpart or similar materials have been previously consumed as food, then the new food or food component should be evaluated on the basis of its own composition and properties.

As an example of the application of substantial equivalence, potatoes have long been part of the human diet. The presence of viral coat proteins in the potato are due to natural viral infections; consequently, these proteins have a long history of human consumption. Coat proteins have never been associated with a toxicity problem and are not considered a food safety issue. Consequently, a potato in which the coat protein of one of these viruses is expressed after the gene has been introduced would be considered substantially equivalent to the infected potatoes that have a long history of safe use and consumption provided the amounts expressed were not grossly different from those occurring following natural infection.

This analogy applies only to viral coat proteins in the portions of the plant traditionally consumed, taking into account the characteristics of the new trait and possible untoward effects of the modification on alkaloid levels and key nutrient starches, as well as the extent of consumption. Some specific examples of additional considerations which it may be necessary to take into account when applying the concept of substantial equivalence are indicated in the following paragraphs. The intended use(s) and degree of exposure must also be considered in assessing safety. This includes the effect(s) of the level of the food or food component in the diet, the pattern of dietary consumption, and the characteristics of the consuming populations (i.e. infants, the elderly, the immunocompromised, etc.).

The consideration of safety may include the need to evaluate possible effects occurring through cooking or other processing. For example, trypsin inhibitors from certain leguminous plants, such as the cowpea trypsin inhibitor, have a long history of safe consumption when properly cooked. However, if the cowpea trypsin inhibitor is expressed in other plants, the safety question relates to whether the normal use of these plants as food involves cooking sufficient for its inactivation. In special cases, depending on the product consumed, the consideration of safety may also include the need to evaluate the potential for, and human health implications of, transfer of the new genetic material.

For example, the use of some antibiotic resistance markers in micro-organisms should be carefully considered since transfer to the microflora of the human gut could, if demonstrated, possibly have human health implications. Another consideration is the influence of the newly introduced modifications on the nutritional value of the food or food component(s). For the majority of modifications being carried out, such changes are unlikely. Nonetheless, when modifications are directed at metabolic pathways of key macro or micro nutrients, the possibility of an impact on nutritional value is increased. Such impacts are of potential significance in cases where the modified food or food component may become a major dietary source of the nutrient affected.

10

Safety of Genetically Modified Food Products

A Joint FAO/WHO Expert Consultation on Foods Derived from Biotechnology was held at the Headquarters of the World Health Organisation (WHO) in Geneva from 29 May to 2 June 2000. The expert Consultations convened by FAO/WHO and OECD have recommended that substantial equivalence be an important component in the safety assessment of foods and food ingredients derived from genetically modified plants intended for human consumption.

This concept embodies a science-based approach in which a genetically modified food is compared to its existing, appropriate counterpart. The approach is not intended to establish absolute safety, which is an unattainable goal for any food. Rather, the goal of this approach is to ensure that the food, and any substances that have been introduced into the food as a result of genetic modification, is as safe as its traditional counterpart.

Several countries have used the concept of substantial equivalence as an important component of the safety evaluations of foods and food ingredients derived from genetically modified plants. They have found this approach to be scientifically sound and practical. Nevertheless, there has not been a universal consensus on the application of this concept. This has resulted in criticism that the approach does not provide a sufficient basis for safety and calls for national governments and international bodies to consider alternative approaches.

FAO and WHO convened this Consultation to evaluate experience gained since the 1996 Joint FAO/WHO Consultation and to assess whether any new scientific information would suggest a need for modifying current approaches for assessing the safety of foods and food ingredients derived from genetically modified plants. This Consultation also provided an opportunity, in the light of recent scientific reports, to review the scientific basis, application, and limitations of the concept of substantial equivalence.

The Consultation was convened to address food safety and nutritional questions regarding foods and food ingredients derived from plants that have been genetically modified using recombinant DNA techniques. Specifically, the Consultation was requested:

— to provide FAO, WHO and their Member States with scientific support in relation to the safety and nutritional features of foods derived from biotechnology on the basis of available scientific data, taking into consideration work done by national authorities, FAO, WHO and other international organisations and other relevant international fora;

— to review existing strategies for the safety and nutritional assessment of foods derived from biotechnology, taking into account ever increasing public concerns and experiences accumulated in testing such foods;

— to make recommendations on further research needs and priorities for evaluation of safety and nutritional aspects of foods derived from biotechnology.

The first session of the Codex ad hoc Intergovernmental Task Force on Foods Derived from Biotechnology, which had been established by the twenty-third session of the Codex Alimentarius Commission in June/July 1999, was held in Chiba, Japan in March 2000. It welcomed the initiative of FAO and WHO to convene the Consultation. It identified the following questions on which the Consultation was invited to formulate scientific opinions:

— What overarching scientific principles should be applied to the safety and nutritional assessment?

— What is the role, and what are the limitations, of substantial equivalence in the safety and nutritional assessment? Are there alternative strategies to substantial equivalence that should be used for the safety and nutritional assessment?

— What scientific approach can be used to monitor and assess possible long-term health effects or unintended/ unexpected adverse effects?

— What scientific approach can be used to assess the potential allergenicity?

— What scientific approach can be used to assess the possible risks arising from the use of antibiotic resistance marker genes in plants and microorganisms?

The Consultation did not consider environmental safety issues related to the release of genetically modified food/ plants into the environment as these were outside its defined scope. Similarly, it did not consider non-science aspects such as socio-economics, risk management and public acceptance.

Food Safety Evaluation of Genetically Modified Foods

For many years the practical difficulties of obtaining meaningful information from conventional toxicology studies on the safety of whole foods have been well recognised. The limitations of conventional toxicological studies became particularly apparent when animal feeding studies were used to assess the safety of irradiated foods. Animal studies are a major element in the safety assessment of many compounds such as pesticides, pharmaceuticals, industrial chemicals and food additives. In most cases however, the test substance is well characterised, of known purity, of no particular nutritional value and human exposure is generally low. It is therefore relatively straightforward to feed such compounds to animals at a range of doses, some several orders of magnitude greater than the expected human exposure levels, in order to identify any potential adverse health effects of importance to humans.

In this way it is possible, in most cases, to determine levels of exposure at which adverse effects are not observed, and so set safe upper limits by the application of appropriate safety factors. By contrast, foods are complex mixtures of compounds characterised by wide variation in composition and nutritional value. Due to their bulk and effect on satiety they can usually only be fed to animals at low multiples of the amounts that might be present in the human diet. In addition, a key factor to consider in conducting animal studies on foods is the nutritional value and balance of the diets used, to try to avoid the induction of adverse effects which are not related directly to the material itself.

Detecting any potential adverse effects and relating these conclusively to an individual characteristic of the food can therefore be extremely difficult. Another consideration in deciding the need for animal studies is whether it is appropriate to subject experimental animals to such a study if it is unlikely to give rise to meaningful information. In practice, very few foods consumed today have been subject to any toxicological studies, yet they are generally accepted as being safe to eat.

In developing a methodology for the safety assessment of new foods, it was essential to establish a benchmark definition of safe food. This was taken up by OECD in 1991 who said that food is considered safe if there is reasonable certainty that no harm will result from its consumption under anticipated conditions of use.

The difficulties of applying traditional toxicological testing and risk assessment procedures to whole foods, meant that an alternative approach was required for the safety assessment of genetically modified foods. This led to development of the concept of substantial equivalence. This approach acknowledges that the goal of the assessment is not establishing absolute safety but to consider whether the genetically modified food is as safe as its traditional counterpart, where such a counterpart exists.

The Consultation agreed that the practical difficulties already identified in relation to the application of conventional toxicology studies to whole food preclude their use as a routine safety assessment technique for genetically modified foods. The Consultation also recognised that the use of toxicology studies could not be justified from an animal welfare perspective where it was unlikely to result in meaningful information. In addition, the Consultation noted that the concept of substantial equivalence was a good example of an approach to reduce the use of animals in toxicology studies by refining safety assessment techniques and replacing animal models with alternatives.

The concept of substantial equivalence was developed as a practical approach to the safety assessment of genetically modified foods. The Consultation agreed that substantial equivalence should be seen as a key step in the safety assessment process. The application of the concept is not a safety assessment in itself; it does not characterise the hasard, rather it is used to structure the safety assessment of a genetically modified food relative to its conventional counterpart. As a starting point, the genetically modified organism (plant, micro-organism or animal), and/or foods derived from it, is compared with its closest traditional counterpart in order to identify any intended and unintended differences which then become the focus of the safety assessment.

Data for comparison should be obtained using validated methods and analysed using appropriate statistical techniques. The comparative approach should take into account agronomic, genetic and chemical aspects and only when all of these have been considered can an objective assessment of safety be made. The type and extent of further studies depend on the nature of the differences and whether or not they are well characterised. Studies should be carried out in accordance with good laboratory practise.

Safety Assessment of Foods

Several international organisations have already addressed the issues associated with the safety assessment of novel foods and, in the present context, genetically modified plants and micro-organisms. It is generally agreed that such an assessment requires an integrated and stepwise, case-by-case approach and some authorities have developed decision trees to assist in determining the extent of testing required in specific cases. This approach is useful in determining appropriate safety assessment strategies.

The safety assessment of a genetically modified food is directed by the results of a comparison between the genetically modified food and its conventional counterpart. It follows a stepwise process aided by a series of structured questions. Factors taken into account in the safety assessment include:

— identity;

— source;

— composition;

— effects of processing/cooking;

— transformation process;

— the recombinant DNA (e.g. stability of insertion, potential for gene transfer);

— protein expression product of the novel DNA;

  — effects on function;

  — potential toxicity;

  — potential allergenicity;

— possible secondary effects from gene expression or the disruption of the host DNA or metabolic pathways, including composition of critical macro-, micro-nutrients, anti-nutrients, endogenous toxicants, allergens, and physiologically active substances; and,

— potential intake and dietary impact of the introduction of the genetically modified food.

The above factors are particularly pertinent to the assessment of foods derived from genetically modified plants. When assessing the safety of foods derived from genetically modified animals and micro-organisms, other factors may need to be taken into account on a case-by-case basis.

If the characterisation of the food indicates that the available data are insufficient for a thorough safety assessment, animal testing may be deemed necessary. This would particularly be the case if the food were expected to make a significant dietary contribution, if there is no history of consumption of the novel gene product or if the modification affects several metabolic pathways.

In the situation where the genetically modified food differs from the traditional counterpart by the presence of one or a few new genes and their products, it may be possible to isolate and study these in a manner analogous to conventional toxicity testing of food additives.

However it is essential to ensure that the material tested is biochemically and functionally equivalent to that produced in the genetically modified food. This provides the possibility of increasing the sensitivity of toxicity tests compared with that possible if the products of the genetically modified plants had been fed directly and avoids some of the artefacts that can occur in toxicity tests on whole foods. However, this strategy is only applicable if the preceding detailed analysis does not reveal significant changes other than those expected. Otherwise testing of the whole food may be required.

When animal testing is conducted on the whole food, it should generally be on the food as consumed by humans. The type of animal study would need to be considered on a case by case basis. In addition to investigating potential toxicological effects, animal studies may also be required if the genetic modification directly or indirectly affects the content or bioavailability of nutrients.

Where toxicological studies are considered necessary to assess the safety of long term consumption of a food in the diet, it is generally considered that a sub-chronic study of 90-days duration is the minimum requirement to demonstrate the safety of repeated consumption of a food in the diet. This may need to be preceded by a pilot study of short duration to ensure that the diet is palatable to the test species and that the levels of incorporation of the test article are appropriate, e.g. the control diet containing the equivalent level of the comparator does not produce effects, as a result of normal levels of natural toxicants present in traditional foods accepted as safe. The highest dose level used in any animal study should be the maximum achievable without causing nutritional imbalance while the lowest level used should be comparable to the anticipated human intake.

The need for additional toxicological tests should be considered on a case-by-case basis taking into account the results of the 90-day study and other studies. For example, proliferative changes in tissues during the 90-day study may indicate the need for a longer-term toxicity study. Conventional toxicological tests are of limited value in assessing whole foods, including genetically modified foods. Based on the maximum levels of the whole food that can be incorporated into experimental diets as indicated previously, a margin of safety may be estimated based on the absence or nature of adverse effects and likely human exposure.

Improved experimental designs should take into account the need for nutritionally adequate animal diets, avoiding some of the inappropriate testing of foods or products. It has been suggested that the use of biomarkers of early effects might increase diagnostic value and sensitivity of toxicity tests on foods. However, it will be necessary not to confuse adaptive and toxic effects in applying this approach.

Unintended Effects

In achieving the objective of conferring a specific target

trait (intended effect) to the host organism by the insertion of defined DNA sequences, additional traits could, theoretically, be acquired or existing traits lost (unintended effects). The assessment of genetically modified foods involves methods to detect such unintended effects and procedures to evaluate their biological relevance and their impact on food safety. Unintended effects may be due to factors such as random insertion events which might result in disruption of existing genes, modifications of protein expression or formation of new metabolites.

The expression of enzymes at high levels may give rise to secondary biochemical effects, e.g. an altered metabolic flux resulting in changed metabolite patterns. The potential occurrence of unintended effects is not specific to the use of recombinant DNA techniques. Rather, it is an inherent and general phenomenon that can occur in conventional breeding. One of the approaches adopted to cope with this problem is to select/ discard plants with unusual and undesired phenotypic and agronomic parameters at an early stage of the plant variety development.

The practice of consecutive back-crossing is also a common procedure used to eliminate unintended effects. Unintended effects due to genetic modification may be subdivided into two groups: those which are "predictable" based on metabolic connections to the intended effect or knowledge of the site of insertion and those which are "unexpected".

Due to the increased precision of genetic modification compared to conventional breeding, it may become easier to predict pathways likely to be influenced by unintended effects. The comparator used to detect unintended effects should ideally be the near isogenic parental line grown under identical conditions. In practice, this may not be feasible at all times, in which case a line as close as possible should be chosen.

The resulting natural variation should be taken into account in assessing the statistical significance of the unintended effect. Where statistically significant unintended differences are observed, their biological significance should be assessed. This may be assisted by knowledge of the mechanisms leading to the changes. In order to assess the biological and safety relevance of an unintended effect, data on the genetically modified plant should be compared to data on other conventional varieties and literature data. If the differences exceed natural variations in traditional food crops, further assessment is required.

Present approaches to assess possible unintended effects are based, in part, on the analysis of specific components (targeted approach). In order to increase the probability of detecting unintended effects, profiling techniques are considered as useful alternatives (non-targeted approach). Profiling techniques are used at different level e.g. genomics, proteomics and metabolomics, and may contribute to the detection of differences in a more extensive way than targeted chemical analysis. However, they are not yet fully developed and validated and have certain limitations. In the future, genetic modifications of plants are likely to be more complex perhaps involving multiple between-species transfers and this may lead to an increased chance of 7 unintended effects. Where differences are observed using profiling techniques, the possible implications of the differences with respect to health need to be considered.

Concept of Substantial Equivalence

The Consultation acknowledged that the concept of substantial equivalence had attracted criticism. This criticism relates, in part, to the mistaken perception that the determination of substantial equivalence was the end point of a safety assessment rather than the starting point. Further disagreement may have arisen from reference to three outcomes of substantial equivalence discussed previously (i.e. substantially equivalent, substantially equivalent apart from defined differences, and not substantially equivalent). Having considered the way in which the concept of substantial equivalence is currently used, and the possible use of alternative strategies, the Consultation concluded that application of the substantial equivalence concept contributes to a robust safety assessment framework.

The Consultation was satisfied with the approach used to assess the safety of the genetically modified foods that have been approved for commercial use. It was agreed that communication of the principles involved in safety assessment could be improved. The Consultation concluded that the key message to be conveyed is that substantial equivalence is a concept used to identify similarities and differences between the genetically modified food and a comparator with a history of safe food use which subsequently guides the safety assessment process. The Consultation reiterated that a consideration of compositional changes was not the sole basis for determining safety. Safety can only be determined when the results of all aspects under comparison are integrated.

It was recognised that whole foods do not lend themselves to the standard safety evaluation principles used for food additives and other chemicals and that a quantitative assessment of risk of individual whole foods from whatever source cannot be achieved. The Consultation agreed that assessing safety relative to existing foods offered the best means of assessing the safety of genetically modified foods. The Consultation considered the issue of long term effects from the consumption of genetically modified foods and noted that very little is known about the potential long term effects of any foods. In many cases, this is further confounded by wide genetic variability in the population, such that some individuals may have a greater predisposition to food-related effects.

In this context, the Consultation acknowledged that for genetically modified foods, the pre-marketing safety assessment already gives assurance that the food is as safe as its conventional counterpart. Accordingly it was considered that the possibility of long term effects being specifically attributable to genetically modified foods would be highly unlikely. Furthermore, it was recognised that observational epidemiological studies would be unlikely to identify any such effects against a background of undesirable effects of conventional foods.

Experimental studies, such as randomised controlled trials (RCTs), if properly designed and conducted, could be used to investigate the medium/long term effects of any foods, including genetically modified foods. Such studies could provide additional evidence for human safety, but would be difficult to conduct. In this respect, it is also important to recognise the wide variation in diets and dietary components from day to day and year to year.

The Consultation was of the view that there were presently no alternative strategies that would provide a better assurance of safety for genetically modified foods than the appropriate use of the concept of substantial equivalence. Nevertheless, it was agreed that some aspects of the steps in the safety assessment process could be refined to keep abreast of developments in genetic modification technology. New methodologies, such as profiling techniques, offer the means of providing a more detailed analytical comparison. However, it was recognised that much more developmental work was necessary before such methods could be validated.

Nutrition-Related Issues

The variety of foods consumed by humans has changed greatly over the centuries, altering the balance of nutrients in the diet. Plant breeding by traditional methods, mutations and recombinant DNA technique can be used to alter nutritional quality and functional traits. Thirty years ago, traditional breeding was utilised to identify and select rapeseed plants free of the nutritionally undesirable fatty acid, erucic acid. The resulting plant, canola, produced an oil that has a more desirable fatty acid profile.

Canola oil now makes up a significant proportion of the daily lipid intake of the consumers in most of the developed nations of the world. More recently, mutational breeding has been used to genetically modify flax plants to produce edible oils instead of the traditional high linolenic industrial oil.

The introduction of new foods and the growing interest in functional foods also have the potential to modify the food supply. The availability of recombinant DNA techniques provides the opportunity to develop foods that help optimise health status. The major difference with recombinant DNA techniques is the ability to introduce different nutrient profiles with greater speed and precision.

Genetic modification of plants now under cultivation have been directed towards agronomic enhancement. Nutritional changes may have a more profound impact on the health of the population. At present, there are no foods derived from genetically modified plants modified to enhance nutrition in the commercial market.

However, there are several plants with altered nutrient composition being developed using recombinant DNA technology. These have been designed to modify nutrient composition and levels or change the functionality of a product. An example of the latter is potato tubers containing increased amounts of starch that is distributed more uniformly, resulting in more efficient processing, lower fat absorption and improved texture. It is anticipated that other genetically modified plants will be developed with nutritional characteristics targeting major health problems. There are a number of examples of genetically modified plants from which foods are produced with the objective of improving health and enhancing food functionality.

These include two examples of genetically modified rice varieties, one in which beta-carotene was produced and another in which an undesirable component for sake brewing (glutelin) was decreased. Other examples were oil seeds in which the fatty acid profile was changed either by traditional mutational techniques or recombinant DNA technique. Recently, canola and soy oils with combined low levels of saturated fatty acids and increased oleic acid, have been produced with the objective of lowering total and low density lipoprotein (LDL) cholesterol levels, one of the risk factors for cardiovascular disease, while at the same time enhancing the functionality of the oil. The Consultation heard details of the recently reported "golden rice" which was specifically designed to target Vitamin A deficiency, a cause of blindness among people living in developing countries. These examples highlight the potential of foods with modified nutritional profiles to reduce the incidence of nutrition-related conditions or diseases.

Targeting a Single Nutrient

Traditional plant breeding techniques of intra and inter-species crossing and mutation are designed to create genetic variation upon which selection of the most desired genotype is the expected outcome. All plant breeding procedures can produce unexpected effects. Low glutelin genetically modified rice, created using an anti-sense technique, signifies improvements in rice storage proteins for commercial sake brewing. The decrease in glutelin levels was however associated with an unintended increase in levels of prolamins. This illustrated that a targeted change in the level of a specific protein can result in other proteins being affected.

The change in prolamin levels did not affect the industrial application but could affect nutritional quality and allergenic potential if the rice were used as a food. Other examples were given regarding soybean and rice showing that genetically modified soy with increased lysine showed an unexpected decrease in oil content, and the genetically modified "golden rice" designed to express betacarotene unexpectedly accumulated xanthophylls. The use of tissue-specific promoters was suggested as a means to limit the number and extent of unexpected effects. In the case of the low glutelin rice, the change in prolamins would not be detected by standard nutritional analyses such as total protein and amino acid profiles. This was only observed following sodium dodecyl sulphate (SDS) gel electrophoresis.

The unexpected finding of xanthophylls in the beta-carotene-enhanced genetically modified rice would not have been apparent from standard nutritional analyses. This difference was observed following high pressure liquid chromatography (HPLC) analyses for carotenoids. Thus, it is important that appropriate analyses of nutrients should be carried out to identify unexpected changes. The ability to change nutrient levels in crop plants through plant breeding, including the use of recombinant DNA techniques, has the potential to result in broad changes in at least two ways: (1) the intended modification in plant constituents could change the overall nutrient profile of the plant product and this change could affect nutritional status of the individual, (2) in addition, unexpected alterations in nutrients could also affect nutrient profiles of the product and nutritional status of people.

Although the genetically modified plant components may be assessed as safe individually, the impact of the change on the overall nutrient profile must be determined. Because changes in individual nutrients could affect a number of plant processes and nutritional outcomes, it is recommended that integration of nutritional and toxicological expertise needed for the evaluation of genetically modified foods be encouraged and facilitated. Consideration should be given to assessing the potential health impact resulting from cnanges in nutrient profile arising from all types of plant breeding.

Examples of the use of mutational breeding to alter the nutritional characteristics of plants include the modification of flaxseed oil from a high linolenic industrial oil to a high linolenic oil similar to corn oil in its fatty acid composition, and genetically modified soy and canola plants being developed that produce a high oleic acid (80-90%) oil that also displays very low levels of saturated and polyunsaturated fatty acids. It will be important to determine if the overall nutrient profile of a genetically modified food has been changed and if dietary intake patterns are altered by the introduction of foods from genetically modified plants.

The introduction of a significant nutritional change in a food may require post-market assessment to determine whether the overall diet has been altered and to what degree, before an assessment of the impact on nutritional status can be made. It is important to ascertain to what extent the modified nutrient is bioavailable and remains stable with time, processing, and storage. For example, the question was raised as to what extent carotenoids in the genetically modified rice remained stable under storage conditions encountered in the developing countries.

Assurance of Safety

Where additional assurance of safety is sought, analytical methods traditionally applied in the evaluation of food constituents such as total protein, fat, ash, fibre and micronutrients may need to be augmented with additional analyses to identify unexpected effects and altered nutrient profiles and bioavailability which may impact on dietary intake and health. Because of the potential for broad changes in nutrient levels and interactions with other nutrients and unexpected effects, it may be necessary in certain instances to undertake feeding studies in animals to determine outcomes that result from changes in nutrient profiles and nutrient bioavailability. Nutritional modifications which are within the normal range of nutrient variation might require a less extensive evaluation than those outside normal ranges.

Genetically modified foods have the potential to improve the nutritional status of individuals and provide products with enhanced functionality for populations in developed and developing countries. The major issues relate to possible nutritional imbalances and the introduction of unexpected alterations in nutrients and other compounds. The change in nutrient levels in a particular crop plant may impact overall dietary intake. In such cases, it would be important to monitor changes in nutrient levels and bioavailability in such foods and evaluate their potential effect on nutritional and health status of consumers. However, an assessment of the impact on nutritional status of consumers is important for all significant dietary changes and not specific to the introduction of genetically modified foods.

Specific Food Safety Issues

This section deals with specific issues that are frequently raised with regard to the safety of genetically modified foods. These issues include the potential for gene transfer from genetically modified plants to gut microflora and mammalian cells, the safety of antibiotic resistance genes as markers for the selection of genetically modified plants, and the assessment of the potential allergenicity of genetically modified foods that may be caused by the presence in these foods of novel gene products. The discussion which follows provides an evaluation of existing knowledge about these topics and elaborates scientific approaches that may be used to assess possible health risks.

Gene Transfer from Genetically Modified Plants: Mechanisms and Consequences for Food Safety

As background to the discussion that follows, it should be noted that all foods contain DNA, which is ingested in significant quantities. In humans, dietary intakes of RNA and DNA vary widely but are typically in the range from 0.1 to 1.0 g per day. Any concerns over the presence of novel DNA in a genetically modified food consumed in the human diet must take into consideration that this DNA would represent less than 1/250,000 of the total amount of DNA consumed. In view of this and the digestibility of dietary DNA, the probability of transfer of genes from genetically modified plants to mammalian cells is extremely low. It is nevertheless necessary to examine this possibility and the consequences of such transfer if it were to occur.

The transfer of plant DNA into microbial or mammalian cells under normal circumstances of dietary exposure would require all of the following events to occur:

— the relevant gene(s) in the plant DNA would have to be released, probably as linear fragments;

— the gene(s) would have to survive nucleases in the plant and in the gastrointestinal tract;

— the gene(s) would have to compete for uptake with dietary DNA;

— the recipient bacteria or mammalian cells would have to be competent for transformation and the gene(s) would have to survive their restriction enzymes; and

— the gene(s) would have to be inserted into the host DNA by rare repair or recombination events.

There have been numerous experiments aimed at evaluating the possibility of transfer of plant DNA to microbes and mammalian cells. To date, there are no reports that marker genes in plant DNA transfer to these cells. Results of model experiments in which mice were orally administered high doses of bacterially derived DNA indicated apparent incorporation of the test DNA fragments into bacterial and mouse cells. The significance of the observations of Schubbert et al. have been seriously questioned. It was concluded that the data do not demonstrate that plant DNA can be transferred to and stably maintained in mammalian cells.

There is additionally no evidence that intact genes from plants can be transferred to and be expressed in mammalian cells. It should be noted that the vast majority of known bacteria are not naturally transformable and there is as yet no evidence of transfer to and expression of plant genes in bacteria under natural conditions. Transfer has been observed under laboratory conditions, but only if homologous recombination is possible. It should be noted that inserted gene sequences in genetically modified plants show in many cases homology with prokaryotic genes. The Consultation is aware of the study being undertaken whereby chickens and sheep are being fed genetically modified maize, and bacteria in the normal flora of the gastrointestinal tract are being tested for DNA uptake.

Should horizontal gene transfer from a genetically modified plant to bacteria occur, the gene (e.g. an antibiotic resistance gene) may alter the fitness of the recipient cell. A reduction in fitness may not provide sufficient selective pressure to eliminate the gene or gene fragment from the gene pool. The presence of this DNA in the cell population could then serve as a genetic reserve for the evolution of the recipient species. The available knowledge on bacteria is derived from bacteria that can be cultured and readily analyzed. Bacteria that are susceptible neither to culture nor identification represent a significant proportion of existing microflora. Therefore, without available knowledge of these bacteria, it is not possible to assess the possibility, probability or consequences of their acquisition of genes or gene fragments.

The consequences of uptake of plant DNA by mammalian cells are different from those of uptake by bacteria because existing data indicate that such DNA is not transmitted via the germline. The extent to which cells containing exogenous DNA are phagocytosed is not yet clear. Neither is it clear that the incorporated DNA is stably maintained and replicated in somatic cells. Mammalian cells would be similarly affected by uptake of exogenously derived DNA regardless of its source.

The most important consideration with respect to horizontal gene transfer is the consequence of a gene being transferred from a genetically modified plant and expressed in recipient cells. Therefore data on the possible extent of such transfer will be needed as part of the safety assessment when the nature of the transferred gene(s) is such that, if transfer were to occur, it would give rise to health concerns.

The Consultation noted that the antibiotic resistance markers currently used in genetically modified plants have been previously reviewed for safety. There is no evidence that the markers currently in use pose a health risk to humans or domestic animals. Nevertheless, with the variable gene transfer frequencies noted in current literature, the transfer and expression of a functional antibiotic resistance gene to recipient cells, while remote, cannot be ignored.

If the recipient cell is subjected to selection from therapeutic use of the antibiotic, proliferation of a drug resistant cell population could compromise the effectiveness of the drug. This directs attention to the more important considerations: whether there are already prevailing high levels of culturable bacteria resistant to that antibiotic, whether that antibiotic is, or could be, clinically important, and whether there are alternative effective therapies.

For certain antibiotic resistance genes currently in use in genetically modified plants, the available data suggest that the consequences of horizontal gene transfer will be unlikely to pose a significant threat to the current therapeutic use of the respective drugs. With other genes that confer resistance to drugs that are important in specific medical use, or to drugs that have limited alternative therapies, the possibility of transfer and expression of these genes is a risk that warrants their avoidance in the genomes of widely disseminated genetically modified organisms and foods and food ingredients.

A number of methods are available to genetically modify plants without incorporation of an antibiotic resistance gene in the commercial product. These methods include removing the gene after successful gene transfer, or using alternative marker genes for genetic transformation. If alternative marker genes are used, they also would need to be evaluated for safety.

In addition, it is recognised that further technical development of these or additional methods may be necessary for practical transformation of certain plant species. In future developments, the Consultation encourages the use of alternative transformation technologies, if available and demonstrated to be safe, that do not result in antibiotic resistance genes in genetically modified foods. If further development of alternative technologies is required, additional research should be strongly encouraged.

Allergenicity

Food allergies are adverse reactions to an otherwise harmless food or food component that involves an abnormal response of the body's immune system to specific protein(s) in foods. True food allergies may involve several types of immunological responses. The most common type of food allergies are mediated by allergen-specific immunoglobulin E (IgE) antibodies. IgE-mediated reactions are known as immediate hypersensitivity reactions because symptoms occur within minutes to a few hours after ingestion of the offending food. IgE-mediated reactions can occur to pollens, mould spores, animal danders, insect venoms and other environmental stimuli as well as foods.

IgE-mediated reactions affect perhaps 10-25% of the population in developed countries, although food allergies represent a small fraction of all allergic diseases. IgE-mediated food allergies affect less than 2.5% of the population in developed countries. Infants and young children are more commonly affected by IgE-mediated food allergies than adults; the prevalence among infants under the age of 3 may be as high as 5-8%. True food allergies also include cell-mediated reactions which involve sensitised tissue-bound lymphocytes rather than antibodies. In cell-mediated reactions, the onset of symptoms occurs more than 8 hours after ingestion of the offending food.

The role of foods in cell-mediated reactions remains uncertain but, celiac disease, also known as glutensensitive enteropathy, affects 1 in every 300 to 3000 individuals in the population depending upon the specific geographic region. The Codex Alimentarius Commission has adopted a list of the most common allergenic foods associated with IgE-mediated reactions on a world-wide basis that includes peanuts, soybeans, milk, eggs, fish, crustacea, wheat, and tree nuts. These commonly allergenic foods account for over 90% of all moderate to severe allergic reactions to foods, although an extensive literature search has revealed more that 160 foods associated with sporadic allergic reactions.

Allergic reactions to fresh fruits and vegetables, the so-called oral allergy syndrome, are also rather common, but these foods are not included on the Codex Alimentarius Commission list because the symptoms are typically mild and confined to the oropharyngeal region and the allergens are unstable to heating and digestion. The list established by the Codex Alimentarius Commission also includes gluten-containing cereals (wheat, rye, barley, oats and spelt) that are implicated in the etiology of gluten-sensitive enteropathy.

The symptoms of IgE-mediated food allergies can range from mild to severe and lifethreatening. Individuals display different thresholds for the offending food, but the most sensitive food-allergic individuals will experience reactions from exposure to trace quantities of the offending food. Life-threatening reactions usually involve exposure to larger doses of the offending food.

Gluten-sensitive enteropathy is a mal-absorbtion syndrome characterised by body wasting, anaemia, diarrhoea, and bone pain along with other symptoms. The threshold dose needed to provoke the symptoms of gluten-sensitive enteropathy are unknown, but also thought to be quite low. Both IgE-mediated food allergies and gluten-sensitive enteropathy are treated with specific avoidance diets. Since in both cases, the threshold dose is quite low, great care must be taken in the construction of safe and effective avoidance diets. Almost all food allergens are proteins, although the possibility exists that other food components may act as haptens.

Similarly, prolamin proteins from wheat, rye, barley, etc. are involved in the elicitation of gluten-sensitive enteropathy. While the crops from which staple foods are derived contain ten of thousands of different proteins, relatively few are allergenic. The distribution of these proteins varies in different parts of the plant and can be influenced by environmental factors such as climate and disease stress. Conventional breeding introduces additional protein diversity into the food supply. However, variations in the protein composition of our diets brought about through conventional crop improvement practices have had little, if any, effect on the allergenic potential of our major foods.

In contrast, altered dietary preferences can have significant implications for the development of food allergies. For example, allergy to peanut (groundnut) occurs at a significant frequency in North America and Western Europe but not in other countries where peanuts are less commonly eaten. Also, recent food introductions such as kiwi fruit have proven to be additional sources of food allergens. These observations provide confidence that there are not a large number of potential allergens in the food supply, but show that new allergenic foods are sometimes introduced into the marketplace. Because of the above, a clear need exists to pay particular attention to allergenicity when assessing the safety of foods produced through genetic modification.

Potential Allergenicity of Novel proteins in genetically modified foods

In 1996, the International Food Biotechnology Council and the Allergy and Immunology Institute of the International Life Sciences Institute developed a decision-tree approach. This allergy assessment strategy has been widely adopted by the agricultural biotechnology industry. This strategy focuses on the source of the gene, the sequence homology of the newly introduced protein to known allergens, the immunochemical binding of the newly introduced protein with IgE from the blood serum of individuals with known allergies to the source of the transferred genetic material, and the physicochemical properties of the newly introduced protein. This study has been adapted by the Consultation for the assessment of the allergenicity of genetically-modified foods [Figure 1].

Since genetically modified foods usually contain novel proteins, their safety should include an assessment of the allergenicity of such novel proteins. The current decision-tree approach requires the examination of a number of parameters which are common to many food allergens. These characteristics facilitate the identification of potentially allergenic gene products, although no single criterion is sufficient to confirm allergenicity or the lack thereof. The relevant criteria used in the current decision tree include:

— *Source of the transferred genetic material*: Particular caution must be exercised if the source of this material

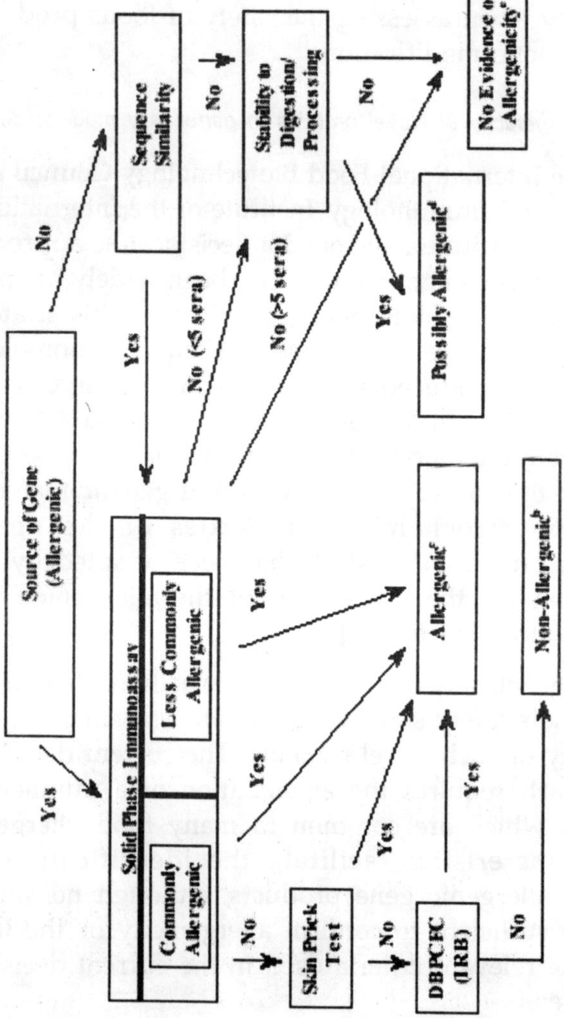

Figure 1. Assessment of the allergenic potential of foods derived From genetically modified crop plants

contains known allergens.

— *Sequence homology*: The amino acid sequence of many allergens is readily available.

— *Immunoreactivity of the newly introduced protein*: If the novel protein is derived from a known allergenic source or if it has sequence homology with a known allergen, then the reactivity of this novel protein with IgE from the blood serum of appropriate allergic individuals is determined.

— *Effect of pH and/or digestion*: Most allergens are resistant to gastric acidity and to digestive proteases.

— *Heat or processing stability*: Labile allergens in foods that are eaten cooked or undergo other processing before consumption are of less concern.

The desirability of including other relevant criteria to improve the reliability of the allergy assessment decision-tree approach was discussed. When the genetically modified food contains genes selected from sources with known allergenic effects, then it must be assumed that the novel gene product is allergenic unless proven otherwise. The current decision-tree approach which advocates the assessment of the binding of the novel protein with IgE from the blood serum of individuals who are allergic to the source of the donor genetic material followed, if necessary by skin testing and blinded oral food challenges, was considered adequate and essential.

The assessment of any unintended effects on the allergenicity of the host material after a genetic modification with genes from other sources, whether allergenic or not, was not considered necessary except in circumstances where the genetic modification could be predicted to alter the protein content of the host product significantly. When the genetically modified food contain genes from sources with no history of allergenicity, the current decision-tree approach relies primarily upon sequence homology comparisons to known allergens and the stability of the novel protein to digestion and processing. It is widely recognised that these two criteria alone may not be sufficient to assess the potential allergenicity of genetically modified foods containing genes from sources with no history of allergenicity.

The current criteria used to determine significant sequence similarity, a match of at least eight contiguous, identical amino acids has been criticised. Suggestions have been made that sequence similarity should instead require a match of a smaller number of contiguous, identical amino acids, perhaps as few as four amino acids. The use of a match of eight contiguous, identical amino acids appears to have some relevance based upon the minimum peptide length for a T cell-binding epitope. Also, it is recognised that the criterion cannot identify discontinuous or conformational epitopes that depend upon the tertiary structure of the protein.

However, the stability of food allergens to heat processing argues for greater significance of linear, continuous epitopes for these particular allergens. International scientific consensus should be sought on the use of sequence homology in the assessment of the allergenicity of genetically modified foods. The use of digestive stability appears to be a rather useful criterion in the assessment of the allergenicity of genetically modified foods. Simulated gastric and intestinal digestive models of mammalian digestion have been used to assess the digestive stability of known food allergens and proteins introduced into foods through genetic modification. While the usefulness of this criterion is apparent, consensus is needed on the ideal protocols for assessment of digestive stability. It is recognised that novel proteins may exist that are stable to digestion but will not become allergens.

Additional testing is needed to assess the allergenic potential of such proteins. The desirability of the development of additional tests to assess the allergenic potential of foods containing genes from sources with no history of allergenicity has been widely expressed. Two additional tests seem to show some promise for addition to the decision-tree approach. The level and site of expression of the novel protein is an important component of the assessment of allergenicity. Novel proteins expressed at comparatively low amounts in the food would have limited potential for allergic sensitisation. Major food allergens are usually major proteins in commonly consumed foods. Thus, greater scrutiny should occur with genetically modified foods containing novel proteins at significant levels in the product.

New proteins expressed in non-edible portions of plants are not a concern in terms of food allergy. Consideration of the function of the novel protein should also form part of the decisiontree assessment of allergenic potential. Certain classes of proteins are well known allergens. For example, the 2S, high-methionine albumins from Brazil nut, walnut, sunflower seed and mustard are major allergens from those sources. Thus, other 2S, high-methionine albumins should be scrutinised very carefully for allergenic potential. Many of the pathogenesis-related proteins of plants display allergenic activity, therefore, the entire class of proteins would also merit close examination. International consensus should be sought on a list of functional proteins with allergenic potential. Certainly, other proteins not on such a list must also be evaluated but this aspect could be a useful part of an overall assessment strategy.

The potential for the use of animal models for the

assessment of the allergenicity of genetically modified foods was also discussed. Unfortunately, reliable, well validated animal models for the assessment of the allergenicity of genetically modified foods do not presently exist, although further research on the development of animal models is encouraged. Other attributes, such as molecular weight and degree of glycosylation, were also discussed. However, it was felt that these attributes were not sufficiently discriminatory to be very helpful.

The novel proteins present in genetically modified foods should also be evaluated for any possible role in the elicitation of gluten-sensitive enteropathy. Clearly, if the desired gene is obtained from wheat, rye, barley, oats, or related cereal grains, the possible role of the novel protein in provoking gluten-sensitive enteropathy must be carefully considered. Furthermore, if genetic modifications are conducted on these cereal grains, possible unintended effects on the gluten proteins should be considered. International consensus should be sought on an appropriate decision-tree approach to the assessment of the role of genetically modified foods and their novel proteins in gluten-sensitive enteropathy.

11

Security Guidelines for Food Processors

Food Security Plan Management

A food security management team and a food security management coordinator should be identified for each plant or company. Each member should be assigned clear responsibilities. A food security plan using established risk management principles should be developed and implemented. The plan should include procedures for handling threats and actual cases of product tampering and an evacuation plan for each facility. Corrective action taken in all cases of product tampering should ensure that adulterated or potentially injurious products do not enter commerce.

The plan should include the immediate recall of adulterated products from trade and consumer channels. Safe handling and disposal of products contaminated with chemical or biological agents should also be included in the plan. A relationship should be established with appropriate analytical laboratories for possible assistance in the investigation of product-tampering cases. Procedures for notifying appropriate law enforcement and public health officials when a food security threat is received, or when evidence of actual product tampering is observed, should be detailed in the plan.

Specially designated entry points for emergency personnel should be identified in the plan. Current local, State and Federal Government Homeland Security contacts and public health officials should be listed in the plan. Members of the food security management team should be trained in all provisions of the plan. Drills should be conducted periodically. The plan should be periodically reviewed and revised as needed. Food security inspections of the facility should be conducted regularly by plant officials to verify key provisions of the plan.

All employees should be encouraged to report any sign of possible product tampering or break in the food security system. Consider implementing an award system or establishing performance standards related to food security consciousness. All threats and incidents of intentional product tampering should be immediately investigated and reported to the local law enforcement officials and the FSIS/State Inspector in-Charge. Liaison with local Homeland Security officials and other law enforcement officials should be pre-established by the food security management team.

Outside Security

Plant boundaries should be secured to prevent unauthorized entry. "No Trespassing" signs should be posted. Integrity of the plant perimeter should be monitored for signs of suspicious activity or unauthorized entry. Outside lighting should be sufficient to allow detection of unusual activities. All access points into the establishment should be secured by guards, alarms, cameras or other security hardware, consistent with national and local fire and safety codes. Emergency exits should be alarmed and have self-locking doors that can be opened only from the inside.

Doors, windows, roof openings, vent openings, trailer bodies, railcars and bulk storage tanks should be secured (e.g., locks, seals, sensors) at all times. Outside storage tanks for hazardous materials and potable water supply should be protected from, and monitored for, unauthorized access. An updated list of plant personnel with open or restricted access to the establishment should be maintained at the security office. Entry into establishments should be controlled by requiring positive identification (e.g., picture IDs, sign-in and sign-out at security or reception, etc.).

Incoming and outgoing vehicles (both private and commercial) should be inspected for unusual cargo or activity. Parking areas for visitors or guests should be situated at a safe distance from the main facility. Vehicles of authorized visitors, guests and employees should be clearly marked (placards, decals, etc.). Truck deliveries should be verified against a roster of scheduled deliveries. Unscheduled deliveries should be held outside the plant premises, if possible, pending verification of shipper and cargo.

Inside Security

Restricted areas inside the plant should be clearly marked and secured. Access to central controls for airflow, water systems, electricity and gas should be restricted and controlled. Updated plant layout schematics should be available at strategic and secured locations in the plant.. Airflow systems should include a provision for immediate isolation of contaminated areas or rooms. Emergency alert systems should be fully operational and tested, and locations of controls should be clearly marked.

Access to in-plant laboratory facilities should be strictly controlled. Comprehensive and validated security and disposal procedures should be in place, particularly for the control of reagents, hazardous materials and live cultures of pathogenic bacteria. Visitors, guests and other non-plant employees (contractors, salespeople, truck drivers, etc.) should be restricted to nonproduct areas unless accompanied by an authorized plant representative. Computer data systems should be protected using passwords, network firewalls and effective and current virus detection systems.

Slaughter and Processing Security

Procedures should be in place to monitor the operation of pieces of equipment (blenders, choppers, poultry chill tanks, etc.) to prevent product tampering. A programme should be in place to ensure the timely identification, segregation and security of all products involved in the event of deliberate product contamination. A validated procedure should be in place to ensure the traceback and trace-forward of all raw materials and finished products.

Projected and actual use of restricted ingredients should be verified at the end of each day, preferably by someone other than the employee who logs the ingredient. Returned goods should be examined for evidence of possible tampering before salvage or use in rework. Records should be kept on the use of all returned goods in rework. The integrity of packaging materials of all spices and restricted ingredients (including premixes prepared in the plant) should be verified before use.

Accurate inventory of finished products should be maintained to allow detection of unexplained additions to or withdrawals from existing stock. Access to product production or holding areas should be restricted to plant employees and FSIS inspection personnel only. Plants should use a system that ensures clear identification of personnel to their specific functions (e.g., colored garb). An updated daily or shift roster of plant personnel should be maintained and distributed to plant supervisors.

Storage Security

Controlled access should be maintained for all product and ingredient storage areas. An access log may be maintained. Security inspection of all storage facilities (including temporary storage vehicles) should be performed regularly, and the results logged. A daily inventory of hazardous chemicals or other products should be made, and all discrepancies should be investigated immediately. Hazardous chemical storage areas or rooms should be secured and isolated from other parts of the plant. In addition, they should be constructed and safely vented in accordance with national or local building codes.

Shipping and Receiving Security

All outgoing shipments should be sealed with tamper-proof, numbered seals that are included on the shipping documents. Establishments should require that incoming shipments be sealed with tamper-proof, numbered seals, and that the seal numbers be shown on the shipping documents for verification prior to entry to the plant. Shipping documents with suspicious alterations should be thoroughly investigated.

All trailers on the premises should be locked and sealed when not being loaded or unloaded. A policy for off-hour deliveries should be established to ensure prior notice of such deliveries and require the presence of an authorized individual to verify and receive the shipment. Packaging integrity of all incoming shipments should be examined at the receiving dock for evidence of tampering.

Advance notification (by phone, e-mail, fax) should be required from suppliers for all incoming deliveries. Notification should include pertinent details about the shipment, including the name of the driver. The FSIS Inspector-in-Charge should be notified immediately when animals with unusual behavior and symptoms are received at the plant. Loading docks should be secured to avoid unverified or unauthorized deliveries. The integrity of food security measures should be a significant consideration in the selection of suppliers of meat and nonmeat ingredients, compressed gas, packaging materials and labels.

Water and Ice Supply Security

Outside access to wells, potable water tanks and ice-making equipment should be secured from unauthorized entry. In-plant ice-making equipment and ice storage facilities should have controlled access. Potable and non-potable water lines in food processing areas should be inspected periodically for possible tampering. The plant should arrange for immediate notification by local health officials in the event the potability of the public water supply is compromised.

Mail Handling Security

Mail handling activity should be done in a separate room or facility, away from in-plant food production/processing operations, if possible. Mail handlers should be trained to recognize and handle suspicious pieces of mail using U.S. Post Office guidelines.

Personnel Security

A system of positive identification/recognition of all plant employees should be in place. Procedures should be established for controlled entry of employees into the plant during both working and non-working hours. New hires (seasonal, temporary, permanent, and contract workers) should be subjected to background checks before hiring. Orientation training on security procedures should be given to all plant employees. The plant should establish and enforce a policy on what personal items may and may not be allowed inside the plant and within production areas.

Genetic Engineering and Food-safety

Biotechnology products, including foods derived from genetically engineered crops, are regulated pursuant to a coordinated framework announced in1986 by the White House Office of Science and Technology Policy. Relying on existing federal laws, the coordinated framework assigns lead regulatory responsibility to one federal agency for each category of product use. For example, the Food and Drug Administration (within the Department of Health and Human Services) is the lead regulatory agency for genetically engineered products in the category of "food and food additives" even though the Food Safety and Inspection Service (within the Department of Agriculture) has jurisdiction over meat and poultry products.

Where agency responsibilities or authorities adjoin or overlap under existing laws, the coordinated framework sets out principles for coordinated and cooperative reviews. In the mid 1980s numerous federal agencies had already amassed considerable experience regulating agricultural, pharmaceutical, and other products developed by traditional genetic manipulation techniques such as selective breeding.

The 1938 Federal Food, Drug, and Cosmetic Act

Regarding the safety of all food, including food developed from biotechnology, the key legislation is the 1938 Federal Food, Drug, and Cosmetic Act (hereafter, FDCA or Act). The following provisions of the Act, as amended, are especially significant:

— *Act § 402. Adulterated Foods*: The adulteration of food and the introduction into interstate commerce of adulterated food is prohibited by the Act. Foods are deemed adulterated, for example, if they contain any poisonous or deleterious substance in a quantity that ordinarily renders the food injurious to health. The Act provides criminal sanctions for violation of its prohibited acts, and perhaps more significantly, by criminalising conduct the Act provides a foundation for civil liability. Section 402 is also the statutory basis of FDA's "post-market" authority to remove food from the market that has been found, through experience or otherwise, to be unsafe.

— *Act § 409. (Unsafe) Food Additives*: The addition of an "unsafe" food additive to food, or the introduction into interstate commerce of food with an "unsafe" food additive, is prohibited. Food additives are "unsafe" unless, for example, the additive and its use are in conformity with a federal regulation prescribing the conditions for safe use. Substances that are "generally recognised as safe" (GRAS) by scientists are excluded from the definition of "food additives" and, therefore, cannot be a § 409 (Unsafe) Food Additive. Importantly, Section 409 is the basis for FDA's only "premarket" approval requirements for genetically engineered food or any other food.

— *Act § 343. Misbranded food*: The misbranding of food or introducing misbranded food into interstate commerce is prohibited. Foods are misbranded if, for example, the label is false or misleading.

— *Act § 701. Regulations and Hearings*: General authority to promulgate regulations for the enforcement of the Federal Food, Drug, and Cosmetic Act is delegated to the Secretary of Health and Human Services (the "departmental home" for FDA).

— *Act § 408a. Pesticide Tolerances*: Summary: Foods containing "unsafe" levels of pesticide residues are brought within the meaning of § 402 Adulterated Foods, thus making their sale unlawful. Unsafe levels of residues include those exceeding the "tolerances" established by EPA. The Administrator of EPA is given the authority to issue regulations which establish, modify, or revoke tolerances for particular pesticide residues. Such tolerances must be "safe," meaning generally that "there is a reasonable certainty that no harm will result from aggregate exposure...."

Whole foods, such as fruits, vegetables, and grains, generally are not subject to premarket approval under Act § 409 (Food Additives) because such foods are generally recognised as safe (GRAS). But should a genetically engineered whole food be subject to strict pre-market review and approval by FDA? FDA gained insight into this question when it chose to conduct a pre-market extensive review of the Flavr Savr tomato, utilising an evaluation process open to public comments and a decision process open to public scrutiny.

The rationale for the premarket extensive review was the uncertainty about whether the genetically engineered Flavr Savr tomato was "substantially equivalent" to existing tomatoes, which are recognised as safe. If it was not substantially equivalent to existing tomatoes, its transgenic food components would be a "food additive" (i.e. not GRAS) and the food would be required to undergo pre-market review and approval under § 409. FDA's conclusion was that the Flavr Savr tomato was "substantially equivalent" to its non-genetically engineered tomato counterparts. More importantly, the experience gained in the Flavr Savr considerations contributed to the development of FDA's 1992 Policy Statement regarding foods derived from new plant varieties.

FDA's 1992 Policy Statement

FDA's 1992 policy statement clarified FDA's legal and

regulatory framework for foods derived from new plant varieties, including new varieties developed through genetic engineering. It established a "standard of care" for the developers of new crop varieties regarding the testing necessary to assure that foods arising from the new plant varieties would be as safe as other foods. It provided guidance as to when a new plant variety would trigger the pre-market approval requirements of Sec. 409. And it provided guidance to producers regarding when they should voluntarily consult with FDA regarding various issues.

Under this framework, foods derived from genetically engineered crop varieties are regulated using an approach identical in principle to that applied to foods derived from conventionally developed new varieties. Under FDA's 1992 policy the safety of food and its regulatory status (is pre-market approval required or not?) depends on specific characteristics of the food and its intended use, irrespective of the method by which the plant variety was developed. In other words...

— if a new plant variety (for example, a new Bt variety of corn) is intended to be used as a food, the safety of that new variety is generally determined by examining the likely presence of toxicants or allergens in the food and any changes in nutritional value;

— the mere presence of trans-genetic material (nucleic acids) in the food does not trigger the pre-market FDA review and approval required by Sec. 409 (the genetic material is GRAS);

— but if the new genetic material expresses itself in the food as a new protein, carbohydrate, fatty acid, oil, or other substance that differs significantly from those currently found in existing foods, then (a) the food is not substantially equivalent to foods already on the market, (b) the "new" proteins, etc., will not be GRAS, and (c) the food with "new" proteins, etc., is subject to the pre-market review and approval requirements of Sec. 409 (unsafe food additives).

For example, if the genes in a new plant variety express themselves in food as a novel protein sweetener, that sweetener would trigger the submission of a § 409 food additive petition by the company and mandatory pre-market approval by FDA. But the mere presence of rDNA in the food, by itself, would not trigger the pre-market approval apparatus. A company could easily have questions about food from its new plant variety. For example, is the food substantially equivalent to existing foods and thereby generally recognised safe (GRAS)? Or is pre-market approval required?

The 1992 policy statement provided guidance on when the company should voluntarily consult with FDA on scientific issues, the design of appropriate test protocols, whether a food additive petition under Sec. 409 would be required, and the requirements for labeling. Although the consultations are technically voluntary, they have become part of the standard of car e expected of indu stry and are relevant in determining civil liability in cases involving unsafe food. As a practical matter, the voluntary consultations are tantamount to being mandatory. A list of completed consultations can be found on the World Wide Web.

A helpful explanation of when a company should consult with FDA, and a decision diagram showing the critical points when consultation should occur, can be found in a 1992 issue of SCIENCE. FDA's 1992 policy also addressed labeling of foods derived from new plant varieties, including plants developed by genetic engineering. The FDCA defines the information that must be disclosed in labeling. The Act also requires that all labeling be truthful and not misleading, but it does not require disclosure in labeling of information solely on the basis of consumers' desire to know. FDA requires special labeling if the composition of the GM food differs significantly from its conventional counterpart.

For example, if a food contains a major new sweetener as a result of genetic modification, a new common name or other labeling may be required. Similarly, if a new food contains a protein derived from a food that commonly causes allergic reactions (and the developer cannot demonstrate that the protein is not an allergen), labeling would be necessary to alert sensitive consumers. Regarding the need to label a food just because it is from a genetically engineered plant variety, FDA does not require foods produced from GM crops to be specially labeled. FDA's rationale is that it has no basis to distinguish genetically engineered foods as a class from foods developed through other methods of plant breeding.

Issues in Food-safety

The use of foods derived from genetic engineering has been controversial. There are some issues related to the food safety. These are;

New Technology and a Patchwork of Old Laws: Does the patchwork of older legislation provide an adequate statutory basis for the regulation of new risks associated with products at the cutting edge of technological innovation? The Coordinated Framework relies on a patchwork of laws such as the 1938 Federal Food, Drug, and Cosmetic Act as the statutory basis for biotechnology regulation. Critics argue that these laws were not enacted to regulate biotechnology and have been stretched beyond their original regulatory intent. Others argue that the laws are sufficient to identify the broad agricultural, environmental, and food-safety concerns related to biotechnology and to assign regulatory oversight to appropriate agencies, and that the laws have been amended as necessary to fix outdated provisions. Furthermore, the broad concepts present in the laws provide flexibility to agencies as they promulgate rules and regulations which are the heart of the regulatory effort.

Regulating the Product, not the Process: Should GM foods be subject to a separate regulatory scheme because they are derived from a genetic engineering process, or should they be regulated like all other foods where the focus is on the characteristics of the product? Consistent with the Coordinated Framework, FDA currently focuses on the characteristics of the food product, not the fact that it may have been produced from a plant produced from the process of genetic engineering, in determining how it is regulated. For example, foods derived from Bt corn or Roundup Ready soybeans are not subject to a separate, mandatory regulatory scheme. However, they are subject to a voluntary consultation process. Critics argue that the process of creating plants through genetic engineering makes GM foods inherently different, creates inherently different risks, and should be subject to a strict regulatory scheme applied to all foods derived from genetically engineered plants.

Pre-market Approval and Substantial Equivalence: Should all foods derived from genetically modified plants be subject to mandatory pre-market approval, even food thought to be "substantially equivalent" to its non-GM counterpart? Most foods have not been tested and approved by FDA before coming to market. Under current laws and regulations the only foods subject to FDA pre-market approval are foods containing added substances. Under its 1992 policy statement, FDA does not generally require pre-market approval for genetically modified foods - it views the GM food as GRAS unless there is a significant difference in its proteins, carbohydrates, etc., compared to the food's non-GM counterpart.

Only if there are significant differences, is pre-market approval required. To date, most of the GM foods reviewed by FDA under its voluntary consultation procedures have not been viewed as significantly different, so most GM foods have not been formally reviewed and approved by FDA before entering the market. Critics argue that no GM foods are exactly equivalent to their non-GM counterparts; therefore, all should be subject to the existing premarket approval requirements applying to food additives.

Voluntary Consultations: Should the voluntary consultative procedure described in FDA's 1992 Policy Statement be made mandatory? Although technically voluntary, the threat of civil liability makes the consultations tantamount to mandatory in the eyes of the companies. And FDA believes all companies that have brought genetically engineered foods to market so far have participated in the voluntary consultations. The results of those consultations are available to consumers under the Freedom of Information Act, and FDA has asked the public for advice about how it might make the consultation data available in a more user friendly manner. Critics argue that the voluntary nature of the consultation, on its face, erodes consumer confidence and, in a procedural sense, doesn't provide the kind of "sunshine" on the decision making process that a mandatory consultative process would provide.

Labeling and Allergies: By not labeling all genetically engineered foods, is FDA putting the public at greater risk of ingesting a new allergen and suffering an allergic reaction? FDA does not believe that a GM food is any more likely to cause an allergic reaction than a non-GM food. It notes that about 90 % of all food allergies in the U.S. are caused by cow's milk, eggs, fish and shellfish, tree nuts, wheat, and legumes (especially peanuts and soybeans). Under existing policy, companies must generally tell consumers on the food label when the food contains a gene from one of these common allergy causing foods.

FDA also indicates it has no scientific evidence to indicate that any of the new proteins introduced into food by GM foods will cause allergies. And, in the unlikely circumstance that the GM food does cause allergic reactions, FDA can exercise its post-market authority to remove the food from stores, just as it would with unsafe foods resulting from other remote risks. Critics argue that a new protein in a GM food could theoretically be a new allergen and there is no known test that can assure it is not; therefore, consumers who wish to choose non-GM foods as a way of avoiding a "new allergen" risk can't tell which foods might be genetically engineered.

Consumer Choice and Labeling: Should genetically engineered foods be labeled to allow consumers to choose? The issue is more complex than it would first appear. Should foods containing some threshold of GM ingredients be subject to mandatory labeling or should the labeling policy simply allow (as it currently allows) the food industry to segregate, label as "GM Free," and supply foods that are below an established threshold for GM free? How do we best use the limited amount of "label space" available? How do you label genetically engineered foods without misleading the public?

Getting Maximum Bang for the Regulatory Buck: If FDA were given new funding to improve the safety of food, would it make sense to invest those new dollars in combating the risks of genetically engineered food or combating other food safety risks? Microbial spoilage and food contaminants (substances like lead or dioxins) probably pose much greater risks to the safety of the food supply than genetic engineering. So should new funds be invested in new programmes to combat food spoilage and contamination, or in expanding FDA's capacity to conduct pre-market reviews and approvals of all foods derived from genetically engineered foods?

Narrow Regulatory Mandates: Do the narrow statutory mandates to agencies prevent them from considering ethical and religious dimensions of biotech-related issues? The regulatory jurisdiction of FDA, for example, is statutorily focused on food safety and labeling issues. FDA has no statutory authority to ban GM foods because of ethical or religious-based concerns about genetic engineering. Some who hold such views believe the statutes should empower the agencies to look beyond their current scope of authority. Others believe that such ethical and religious views are entirely proper as a basis for individual actions, e.g.,consumer boycotts. They argue, however, that agencies shouldn't be enforcers of ethics or religion because such a role would raise the issues of "whose ethics" and "whose religion;" instead, such issues should be deferred to the political arena, subject to constitutional limits on the role of government and protections of individual freedoms.

It should be noted that the narrow mandates of biotech regulators and the current labeling void for genetically engineered foods creates a dilemma for those who object to biotechnology on religious or ethical grounds. They are either forced to recast their objections as concerns about health and environmental risks (if they are to have any impact on agency rule-making), or they must resort to public protests and civil disobedience (if they are to otherwise "live" their beliefs). If it were possible, through some resolution of the labeling issue, to empower these people to "live" their beliefs through consumer choice, such a resolution might bring greater clarity to the arguments about health and environmental risks and defuse some of the public protests about genetically engineered foods.

Importance of Food Quality and Safety

Food systems in developing countries are not always as well organised and developed as in the industrialised world. Moreover, problems of growing population, urbanisation, lack of resources to deal with pre- and post-harvest losses in food, and problems of environmental and food hygiene mean that food systems in developing countries continue to be stressed, adversely affecting quality and safety of food supplies. People in developing countries are therefore exposed to a wide range of potential food quality and safety risks.

In 2020, the world population will most likely reach 7.6 billion, an increase of 31% over the mid-1996 population of 5.8 billion. Approximately 98% of the projected population growth over this period will take place in developing countries. It has also been estimated that between the years 1995 and 2020 the developing world's urban population will double, reaching 3.4 billion. This overall increase in population and in the urban population in particular, poses great challenges to food systems.

Intensification of agriculture and animal husbandry; more efficient food handling, processing and distribution systems; introduction of newer technologies including appropriate application of biotechnology will all have to be exploited to increase food availability to meet the needs of growing populations. Some of these practices and technologies may also pose potential problems of food safety and nutritional quality and call for special attention in order to ensure consumer protection. Rapid urbanisation has led urban services to be stretched beyond their limits, resuiting in inadequate supplies of potable water, sewage disposal and other necessary services.

The issue of street foods merits special attention. Recent growth in this sector has been phenomenal with important economic and nutritional implications in the urban context. Street foods are readily accessible and affordable to urban populations, and they provide the energy and nutrient needs of large segments of workers and their families in the cities. Clean and nutritious street foods have a positive impact on food security; low quality and unsafe street foods can have a negative impact. National and local authorities should take cognisance of the potential of this informal sector to improve food security.

In many cases, facilities and training need to be provided for hygienic handling of street foods to assure their safety and quality. It is often said that the poor will consume "anything" to mitigate their hunger. This may or may not be true. To the extent that this phenomenon exists, it only indicates the trade-off which people may face in difficult situations. On the one hand, survival may depend mainly on access to a minimum quantity of food. On the other hand, consumption of food which does not meet minimum safety standards, can also jeopardise survival. Governments must take the necessary steps through national food security policies, systems and programmes to ensure that food quality and safety considerations form an integral part of their food security system.

At present many countries lack comprehensive national food quality and safety regulations. In weighing the gains against the cost of comprehensive food quality and safety standards, countries may conclude that given their social and economic level of development, the cost of certain standards are high relative to the gains, especially if these higher costs have to be borne by the poor themselves. Nevertheless, some developing countries, with FAO technical assistance, have adopted and implemented comprehensive national food quality and safety standards based on the international recommended Codex Alimentarius Commission standards, guidelines and codes of practice.

These countries have immediately benefited from higher levels of investment in the food sector, better acceptance by consumers of higher quality and safer domestically produced raw and processed foods, and greatly improved access to foreign markets for their food exports. Meeting these Codex-based standards has also increased efficiency in food production, processing and distribution, promoted a lower cost domestic supply of good quality and safe foods, reduced food loss problems and greatly increased export earnings.

The essence of all national food laws in industrialised and developing countries alike, is based on the following basic provision, which may be worded differently but has similar intent: "Any person who sells to the prejudice of the purchaser any food which is not of the nature or is not of the substance, or is not of the quality of the food demanded by the purchaser, shall be guilty of an offence". Such legislation establishes the will of the governments to protect their populations from unsafe and adulterated foods. This is achieved through appropriate food control measures based on well- defined food regulations covering quality and safety of food and its honest presentation to the consumer.

Any steps taken by governments to strengthen these activities would significantly help in meeting food security needs and their commitments at the WFS. In all countries the food industry bears the responsibility of meeting food quality and safety regulatory requirements. The food industry encompasses the activities of small-scale farmers and artisanal fisheries, through medium to large-scale producers; food storage; processing; wholesale and retail marketing. Food chains can be as short as from the home garden to the family table or thousands of kilometers long with many intermediaries.

Food preservation, processing and packaging systems can be minimal or highly sophisticated, but assuring food quality and safety in all situations should be a constant. Industry must play its role in assuring food quality and safety through the application of quality assurance and risk-based food safety systems utilising current scientific knowledge. The implementation of such controls throughout production, handling, processing and marketing leads to improved food quality and safety, increased competitiveness; and, reduction in cost of production and wastage. Through national food control systems, governments should provide a supporting infrastructure and assume an advisory and regulatory role.

Food Trade

The value of the world food trade in 1997 was about $ 458 billion, and is increasing every year, thanks to the expanding world economy, liberalisation in food trade, growing consumer demand and developments in food science, technology, transport and communication sectors. Scientific developments have also allowed a better understanding of the nutritional qualities of foods and their health implications. This has led consumers to become more discriminating in food matters and to demand protection from inferior quality and unsafe foods.

Consumers expect that domestic and imported foods will meet basic quality and safety standards and requirements related to food hygiene, labelling and certification, use of food additives, limits for pesticide residues etc. Access by developing countries to food export markets in general, and of the industrialised world in particular, will depend on their capacity to meet the regulatory requirements of importing countries. For most developing countries, agriculture lies at the centre of their economies and food exports are a major source of foreign exchange and income generation for rural and urban workers in agriculture and agro-industrial sectors.

The long-term solution for developing countries to sustain a demand for their products in world markets lies in building up the trust and confidence of importers in the quality and safety of their food supply systems. This requires improvement within national food control systems and within industry food quality and safety programmes. Such efforts will greatly help in increasing the relatively small share of developing countries in the international food trade. To have a better understanding of current food quality and safety problems in international food exports from developing countries, it is useful to look at the import detentions by the United States Food and Drug Administration (US FDA)- the only agency which makes such data public through a monthly import detentions list.

US FDA regulates all foods in the US other than meat and poultry products. The data from FDA detention lists for the period July 1996 to June 1997 for imported food from different regions of the world is given in Table 1. The majority of detentions and rejections of foods from developing countries are not related to highly technical or sophisticated requirements.

At the top of the list stand food hygiene problems represented by contamination of food with insects and rodent filth. Microbiological contamination comes next, followed by failure to comply with US low acid canned food registration requirements, and then labelling. Over 50% of the rejections are attributable to lack of basic food hygiene, and failure to meet labelling requirements. Dealing with these is well within the means of most developing countries and would go a long way in promoting export trade.

Table 1. Number of Contraventions Cited for United States Food and Drug Administration Import Detentions.

Origin Reason for contravention	Africa	Latin America and the Caribbean	Europe	Asia	Total
Food Additives	2 (0.7 %)	57 (1.5 %)	69 (5.8 %)	426 (7.4 %)	554 (5.0 %)
Pesticide residues	0 (0.0)	821 (21.1 %)	20 (1.7 %)	23 (0.4 %)	864 (7.7 %)
Heavy Metals	1 (0.3)	426 (10.9 %)	26 (2.2 %)	84 (1.5 %)	537 (4.8 %)
Mould	19 (6.3 %)	475 (12.2 %)	27 (2.3%)	49 (0.8 %)	570 (5.1 %)
Microbiological contamination	125 (41.3 %)	246 (6.3 %)	159 (13.4 %)	895 (15.5 %)	1425 (12.8 %)
Decomposition	9 (3.0 %)	206 (5.3 %)	7 (0.6 %)	668 (11.5 %)	890 (8.0 %)
Filth	54 (17.8 %)	1253 (32.2 %)	175 (14.8 %)	2037 (35.2 %)	3519 (31.5 %)
Low Acid Canned Food	4 (1.3 %)	142 (3.6 %)	425 (35.9 %)	829 (14.3 %)	1400 (12.5 %)
Labelling	38 (12.5%)	201 (5.2%)	237 (20.0%)	622 (10.8%)	1098 (9.8%)
Other	51 (16.8 %)	68 (1.7 %)	39 (3.3 %)	151 (2.6 %)	309 (2.8 %)
Totals	303 (100 %)	3895 (100 %)	1184 (100 %)	5784 (100 %)	11166 (100 %)

The international trading environment has changed considerably in the light of agreements reached under the auspices of GATT, and the subsequent establishment of the World Trade Organisation (WTO). Two Agreements are of particular interest as they introduce a measure of discipline in international trade and are extremely relevant to food safety and quality issues. These are:

— Agreement on Application of Sanitary and Phytosanitary Measures (SPS).

— Agreement on Technical Barriers to Trade (TBT).

The SPS Agreement relates to protection of human, animal and plant health and life. The Agreement covers all relevant laws, decrees, regulations; testing, inspection certification and approval procedures; packaging and labelling requirements directly related to food safety. Nations are asked to apply only those measures that are based on scientific principles, and only to the extent necessary and not constituting a disguised restriction on international trade. The Agreement encourages use of international standards where they exist and identifies the Codex Alimentarius Commission (CAC) food standards, guidelines and other recommendations as consistent with the provisions of SPS.

Where a WTO member considers that a higher level of sanitary protection than afforded by Codex is necessary, it will have to produce scientific evidence based on valid risk assessment techniques. The TBT Agreement also recognises international standards where they exist. It requires that technical regulations on traditional quality factors, fraudulent practices, packaging, labelling etc. imposed by countries will not be more restrictive on imported products than they are on products produced domestically. Technical measures applied should not create unnecessary hurdles in international trade, have a legitimate purpose and the cost of their implementation should be proportional to the purpose of the measure.

If the proposed measure is considered to violate the provisions of any of the two Agreements, it can be challenged and brought before the WTO dispute settlement mechanism. The recognition of international food standards in international food trade is an important element of the WTO Agreements. This provides a level playing field to developing countries. Prior to these Agreements developing exporting countries had to abide by the regulations of importing countries. Now the situation is somewhat different. Firstly, the trading nations should apply international standards, and secondly, in case of a dispute, countries can use the WTO dispute settling mechanism where scientific evidence will be the main determining factor in settling SPS or TBT food regulatory disputes. It would therefore be in the interest of exporting countries to insist on acceptance and use of international standards.

More than 800 million people, many of them children, are today hungry and malnourished with serious impact on growth and learning capacity of children and the ability of adults to lead fully productive lives. Moreover, most of these people are to be found in those parts of the world where such food as they have is often contaminated or adulterated, thus reducing its nutritional quality and inflicting severe harm to their nutritional well-being and to their household economies. Food-borne diseases are a worldwide problem of great magnitude, both in terms of human suffering and economic costs.

The task of estimating with any accuracy the occurrence of food-borne diseases globally is truly formidable as in most countries it is poorly recorded. It is estimated that almost 70% of the approximate 1.5 billion episodes of diarrhoea that occur in the world annually are directly caused by biological or chemical contamination in foods. Even when such diseases are not fatal, they severely increase the effects of poor diet owing to reduced intake, nutrient losses and mal-absorption, which may lead to mental retardation and physical disabilities.

There is no evidence that evaluated food additives or pesticides used in accordance with international recommendations have been the cause of any harm to humans. There is, however, a risk that the inappropriate use of such chemicals can cause health problems. Plant toxins have also been implicated in food safety problems. An example is Lathyrus Sativus adulteration in certain food grains which has led to food-borne disease outbreaks. Other outbreaks have involved contaminants like lead, mercury, cadmium; admixture of mustard seeds with argemone seeds; adulteration of olive oil with mineral oil.

Marine-biotoxins have also been implicated in several poisoning episodes. Mould growth by-products called mycotoxins are ubiquitous. Some are powerful carcinogens and can also cause other health problems to humans and animals. There are several recorded cases of mycotoxin contamination having led to serious disease outbreaks. Estimation of the economic consequences of unsafe or contaminated food is complex.

It involves consideration of the value of crops and animal products spoiled or destroyed as a result of such contamination, value of rejections/detentions in the export trade, medical treatment costs, loss of output or earnings resulting from morbidity, disability or premature death. It is the last of these economic consequences that is the most difficult to measure, but on a global basis it is probably the single largest element in the entire cost of unsafe food. In addition, the death or disability of the wage-earner may have disastrous consequences for the quality of life of survivors.

Some studies have been carried out to assess the total costs incurred by society as a consequence of food-borne illnesses. A 1977 study shows the total annual loss attributable to Salmonellosis in the Federal Republic of Germany to be DM 240 million, of which 45% represented losses due to human food-borne disease, and the major part of the rest accounting for infections in cattle and poultry. A 1987 study in USA estimating Salmonellosis medical costs and productivity losses gives a figure of US $ 1.4 billion as total costs. The same USA study gives a total loss figure of US$ 256 million from Listeriosis illnesses and deaths. In USA alone, costs for loss in productivity due to seven specific pathogens have been estimated to range between US$ 6.5 billion to US$ 13.3 billion annually.

Analysis of the economic impact of a Staphylococcus aureus outbreak in India, showed that 41% of the total cost of the outbreak was borne by the affected persons which included loss of wages or productivity loss and other expenses. On the basis of the percentage of per capita income, the economic burden on affected people in India was higher than in the case of a similar outbreak in USA. Food supply systems in developing countries are often fragmented involving a multitude of middlemen. This exposes it to various types of fraudulent practices.

These may include simple adulteration of food with something of lesser value or no value at all, or mislabelling the product with the intent of misleading the consumer. Besides the public health impact due to the reduction in the nutrient content of food or food contamination, the consumer is defrauded. Considering that in developing countries, people spend almost 50% of their earnings on food, and among lower-income households this figure may rise to above 70%, the impact of such fraudulent practices can be quite devastating. Food is a good indicator of the state of the environment in which it is produced.

Monitoring of environmental contaminants in food therefore not only assists in ensuring food safety but can also give early warnings about the state of the environment, such as level of heavy metal contamination to enable appropriate action for maintaining its productivity.

Almost all countries have a food control system, howsoever poorly developed, to protect their populations against unsafe, adulterated, or otherwise poor quality food. This also implies that food legislation exists indicating governmental policy towards consumer protection. Why then do large sections of the populations in many developing countries still suffer from the ravages of unsafe, unhygienic or adulterated food, and severe losses in food export trade take place annually due to food not meeting the basic requirements of quality and safety? There are several causes for this unsatisfactory situation which need attention at the level of the food industry and the government. Some of these are listed below:

—Food systems are complex. In the case of developing countries, they are also highly fragmented and predominated by small producers. This has its own socio-economic advantages. However, as large quantities of food pass through a multitude of food handlers and middlemen extending the food production, processing, storage and distribution chain, control is more difficult and there is a greater risk of exposing food to contamination or adulteration. Lack of resources and infrastructure for post-harvest handling, processing and storage leads to severe diminishing of quality and avoidable contamination and food losses.

— Interaction and cooperation between industry and government on food control matters is often lacking. The basic responsibility of industry is to produce and market good quality and safe food that is honestly presented. It is the duty of the government to ensure compliance by industry to national food quality and safety requirements.

— National food control systems suffer from serious inadequacies, including:

a) They are not based on modern scientific and management concepts using compliance policies, risk assessment, HACCP, transparency, and broad-based involvement of industry, trade and consumers.

b) Insufficient involvement of scientific expertise from the academia, industry, consumers to strengthen the scientific basis for food control decision making processes.

c) Lack of suitable facilities such as laboratories.

d) Lack of resources such trained inspectorate and laboratory staff, funding.

e) Inflexibility of the system making it difficult to cope with developments in food science and technology, changing consumer demands, and newer requirements of trade and industry. Institutional obstacles to reforms can be formidable and can create disincentives for development of industry causing serious damage to national economy.

f) Lack of coherence among different governmental activities concerning agriculture, food, trade, industry and health. Lack of coordination to achieve optimal results.

It may be useful to briefly touch upon some of the basic requirements of food control systems in order to assist countries to review their facilities with a view to making them more efficient and effective.

Quality and safety of food have to be ensured throughout the food production, processing, storage and distribution chain. This is a multi-sectoral activity and its objectives cannot be reached without the active cooperation of producers, traders, industry and government and also the involvement of the scientific community. The involvement of various sectors of the economy in the development and operation of the food control system is a pre-requisite for its success. This can be achieved through a well-conceived national food control strategy developed with the support of the various participants.

The strategy clearly lays down the role of governmental agencies, various sectors of economy, and consumers and establishes mechanisms for cooperation, and the means of dealing with newer or emerging challenges in regard to human health and the national economy. It also ensures that available manpower and financial resources are utilised in a co-ordinated manner to achieve optimal results. Development of such a strategy at national level deserves priority.

In many countries, existing food legislation is outdated needs overall review. The FAO/WHO Model Food Law can provide guidance as necessary. This is a model of an enabling law which covers the basic principles without encumbrances like detailed food regulations that can be written and promulgated separately. The advantage is that detailed regulations, which may need periodic review, can easily be amended, if necessary, without going to the national legislature, which would have been necessary if they were also written into the law.

The model food law lays down broad general principles, basic definitions, responsibilities for implementation, inspection and analytical requirements, penalties, powers to make rules etc. Several governments have already revised their laws in conformity with the requirements of the model law. Those who have not done so may consider applicability of the provisions of the model law within their legal system.

A national food control service consists of three main components: food inspectorate; laboratories and analytical staff; and managerial and supervisory staff which not only have the day-to-day management functions but also the required scientific and legal competence. Food inspectors, who are the eyes and ears of the service, should be properly trained. Similarly, the analytical staff has to be fully trained in order to maintain the credibility of the system. FAO/WHO model food law provides for a central advisory committee or body which co-ordinates inputs from different sectors, industry and consumers, and also makes recommendations on food standards and other regulations.

As such a committee plays a key role in the operation of the food control system, many governments have already established such advisory bodies. There is, however, considerable scope for improving their competence, effectiveness and efficiency. In addition, in order to deal with the current international food trade scenario and, to protect the interests of the domestic consumer as well of the national economy, it has been recommended by FAO/WHO that governments also set up National Codex Committees. Many governments have taken action along these lines.

Lack of compliance policies in most of the developing countries is a serious drawback. A compliance policy is an official statement or group of statements that establish specific or general limits to which products, processes or conditions must comply and be, in accordance with relevant laws and regulations. A compliance policy helps in administering, interpreting and assisting in implementing the law. It brings about uniform application of the law, transparency in decisions, provides specific instructions for the agency staff as well as guidance on compliance matters to the industry.

Development of a compliance policy calls for a legal authority, resources, scientific expertise, community support, and a periodic review process. Strengthening food control services requires considerable development in infrastructure. The setting up, equipping and maintaining of laboratories require heavy investment. Capacities of the systems have to be built up periodically and this needs resources. Governments will provide such resources only when they have recognised the economic as well as health implications of food control.

Notably, economic ministries usually have a better record in mobilising resources for adequate development of infrastructure and food control capacities. Assistance may be sought from international agencies like FAO, WHO, World Bank and others for this purpose. However, this will depend upon the priority that national authorities give to food control.

There have been success stories where the strength of one developing country has been matched with the needs of another with mutual benefit for both. The UN system has been encouraging this approach of Technical Cooperation among Developing Countries (TCDC). Food control can benefit from this approach, particularly in manpower development and capacity building. Emergence of a number of regional economic groupings, growing food security needs, and trading interests of many developing, have improved the scope for TCDC. This needs to be encouraged.

The Joint FAO/WHO Codex Alimentarius Commission was set up in 1962 and has the major objectives of protecting the health of consumers and ensuring fair practices in food trade. It is an intergovernmental body engaged in preparing international food standards and other relevant recommendations which promote quality and safety of food.

It brings together scientists, technical experts, government regulators and international consumer and industry organisations. Over the past 37 years it has produced over 200 food standards; nearly 3000 maximum residue limits for pesticides, veterinary drugs, mycotoxins, environmental contaminants; codes of hygienic practices; a general standard for food labelling; a code of ethics for international trade in food and a wide range of guidelines and recommendations for governments and industry.

Codex has proved to be one of the most successful programmes of the specialised agencies of the UN, contributing to international harmonisation in the important area of quality and safety of food. Under the SPS agreement the Codex standards, guidelines and other recommendations have been recognised as bench marks for international harmonisation. Codex standards also serve as the basic texts for resolution of trade disputes. TBT also recognises international standards.

The work done by Codex in the fields of food quality and safety, consumer protection, and issues related to international food trade is so comprehensive and scientifically-based, that the governments of developing countries are well advised to make full use of this in their efforts to improve their food control. Developing countries should be more actively involved in the work of the Commission to ensure that the interests of their consumer and their national economic interests are adequately considered.

For almost three decades, FAO has been the lead international agency providing technical assistance in the area of food control. Such assistance has covered many aspects of the problem, including:

— advice on review and drafting of food laws and regulations; setting up and strengthening of laboratories; assisting in national food contamination monitoring programmes;

— workshops for development of national food control strategies; training and manpower development in various fields;

— assistance in improving industry food quality and safety programmes, strengthening of national food control infrastructures;

— surveys on specific food safety and consumer protection problems like street foods;

— promoting centres of excellence for training; supplying laboratory reference materials;

— quality control and inspection for exports;

— certification systems for food exports; harmonisation of food regulations at regional levels;

— promoting TCDC; and, strengthening national Codex Committees.

The need for assistance is obvious and has in fact grown many-fold as a result of the WTO agreements. The basic problem however, is the paucity of resources. Realising the importance of the subject in terms of food security, it should be possible to have the co-operation and support of financial institutions like World Bank, Regional Development Banks and bilateral donors.

SPS agreement refers to the principle of "equivalence" between one country's protection measures and those of another. It encourages use of mutual recognition agreements between countries. The agreement also asks food-importing countries to provide technical assistance to developing food-exporting countries to enable them to export to those countries. It may be worth considering how to institutionalise and co-ordinate measures for such assistance for the general good of the developing countries and their food control systems in particular.

12

Future of Global Food Production

Irrigated agriculture has been an important contributor to the expansion of national and world food supplies since the 1960s, and is expected to play a major role in feeding the growing world population. However, irrigation accounts for about 72 percent of global and 90 percent of developingcountry water withdrawals, and water availability for irrigation may have to be reduced in many regions in favor of rapidly increasing nonagricultural water uses in industry and households, as well as for environmental purposes. Out of concern over increasing water scarcity for irrigation, the role of water management and investments for irrigated agriculture and food security has received substantial attention in recent years.

However, rainfed areas currently account for 58 percent of world food production. Given the importance of rainfed cereal production, insufficient attention has been paid to the potential of production growth in rainfed areas to play a significant role in meeting future food demand.

Growth of Rainfed Crop Production

In order to increase production, farmers have two options, either to use extensive systems (which expand the area planted) or intensive systems (which increase inputs on a planted area in order to increase yields). In order to meet immediate food demands, farmers in many rainfed areas have expanded production into marginal lands. These fragile areas are susceptible to environmental degradation, particularly erosion, due to intensified farming, grazing and gathering. This problem may be especially severe in areas of Africa, in which the transfer from extensive to intensive systems was slower than in other regions.

Expansion of production into marginal areas can cause many environmental problems. When these areas are cleared, many plants native to the area may be lost, and disease and pest problems may also develop due to changes in the ecosystem. Soil erosion is also often a significant problem in areas of agricultural expansion. Many of the marginal areas to which agriculture expands in the developing world include hillsides and arid areas, which make soil erosion a particular concern. These environmental impacts can lead to additional economic and health problems, particularly for the poor individuals that generally live in marginal areas.

These impacts are generally greater on the poor than on other factions of the population due to the fact that they do not have adequate assets to mitigate the impacts of environmental degradation. Environmental problems can have far-reaching implications in poor communities through decreased agriculture production potential, which may further increase poverty, leading to increased malnutrition and poor health. Increasing production by expanding the planted area into marginal areas may have additional negative impacts on the population that moves into the se areas, as living conditions can be much harsher than in more productive areas.

Because of these environmental consequences of area expansion, crop yield growth is a better solution than increasing the area planted in rainfed areas. Under some circumstances, increasing production on more productive lands-such as irrigated areas-can ease the pressure to use more marginal lands for cropland and help to keep those natural habitats from being destroyed. But as will be seen below, the potential for expansion of irrigated area is limited in most of the world. Therefore, intensive cropping systems that involve increased inputs such as labor, fertilizers, pesticides, or improved varieties to increase yields will be essential for rainfed crop production.

Sustainable intensification of rainfed agriculture development can increase production while limiting environmental impacts. The three primary ways to enhance rainfed agricultural production through higher crop yields are: 1) to increase effective rainfall use through improved water management; 2) to increase crop yields in rainfed areas through agricultural research; and 3) to reform policies and increase investment in rainfed areas.

Water Harvesting

Many developing countries located in arid or semi-arid regions experience significant problems in securing adequate amounts of water for rainfed crop production. Water scarcity problems in arid regions result simply from the lack of sufficient rainfall. Semi-arid regions, however, may receive enough annual rainfall to support crops but it is distributed so unevenly in time or space that rainfed agriculture is not viable. Water loss through evaporation and runoff exacerbates water scarcity problems in these areas. Low rainfall areas that receive between 300 - 600 mm annually may be able tc combat these problems using supplemental irrigation methods, but regions receiving less than 300 mm of annual rainfall must resort to other methods to secure enough water to support crop production.

Water scarcity is a significant problem for farmers in Africa, Asia, and the Near East where 80 - 90 percent of water withdrawals are used for agriculture. While farmers in some high-potential regions have been able to increase yields by 4 - 5 percent in recent years, farmers in the semi-arid tropics of Asia and Africa have only increased agricultural growth by less than 1 percent. Farmers in these arid regions may be particularly hard hit, as development requires more water for domestic and industrial uses. Potential does exist, however, to increase agricultural water use efficiency through water harvesting and conservation techniques.

Some water harvesting methods have proven successful in practice; trials of water harvesting in Burkina Faso, Kenya, Niger, Sudan and Tanzania have shown increased yields of 2 - 3 times those achieved in dryland farming. Water harvesting is a general term usually used to describe the collection and concentration of runoff for many purposes, including agriculture and domestic uses. One characteristic is the importance of storage to many water harvesting systems due to the intermittent water flow in the arid and semi-arid areas where water harvesting takes place. In addition, most water harvesting operations consist of a catchment area and a receiving area for the capture of runoff, and are generally small both in size and in level of investment.

Water harvesting activities occur near the location where the rain falls, therefore the storing of river water in large reservoirs and groundwater mining are generally not included under the category of water harvesting. Water harvesting for agriculture (sometimes referred to as runoff farming) involves concentrating and collecting the rainwater from a larger catchment area onto a smaller cultivated area. The runoff can either be diverted directly and spread on the fields or collected in some way to be used at a later time. Different authors have classified water harvesting methods in various ways.

External catchment rainwater harvesting (sometimes referred to as macrocatchment water harvesting) involves the collection of water from a large area that is a substantial distance from the area where crops are being grown. Types of external catchment systems include runoff farming, which involves collecting sheet or rill runoff from the hillsides into flat areas, and floodwater harvesting within a streambed using barriers to divert stream flow onto an adjacent area, thus increasing infiltration of water into the soil. This type of water harvesting can be used for any number of different crops including row crops, trees or closely growing crops.

Microcatchment water harvesting methods are those in which the catchment area and the cropped area are distinct but adjacent to each other. Some specific microcatchment techniques include contour or semi-circular bunds made of earth, stone or trash, pitting, strip catchment tillage, and a meskat-type system in which the cropped area is immediately below the catchment area that has been stripped of vegetation to increase runoff. These methods are often used for medium water demanding crops such as maize, sorghum, millet and groundnuts.

Rooftop runoff collection involves the collection of runoff from slanted building roofs and is used almost exclusively for domestic consumption. Some other water collection methods that have been used include fog collection and snow collection. Fog collection has been used in some mountainous coastal regions of Central and South America with large amounts of fog. This method utilizes fine nylon net strung between poles, which collects water droplets from condensed fog that is then stored for later use. This method generally does not result in large amounts of water being collected.

Snow harvesting has also been used in some areas of Afghanistan. In this method, snow is collected in the winter and stored in a deep watertight pit, which proceeds to slowly melt over the following summer. This method is not feasible in many arid and semi-arid areas that are located in warmer climates. In situ water harvesting (or water conservation) methods are also used to help increase water use efficiency and are classified as water harvesting.

Many factors influence the usefulness of rainwater harvesting in general as well as the applicability of different methods in a particula r area. Rainfall harvesting is only necessary in arid and semi-arid regions that receive low levels of rainfall or in which there is high intra or inter-seasonal rainfall variability that makes traditional rainfall agriculture infeasible. Rainfall intens ity is one factor that impacts the effectiveness of the chosen water harvesting method.

Other factors that influence the choice of rainwater harvesting method include topography, soil characteristics - particularly those related to water infiltration, and the choice of crop to be planted. Specific topographic characteristics that are necessary for rainwater harvesting include a landscape surface that facilitates runoff, and varia tions in altitude such that runoff flows down the slope and collects at a flat portion of the landscape. In addition, the cropped area soil must be deep enough and of a suitable texture to induce rainfall infiltration and retention.

Due to the unique soil and landscape characteristics across regions, the same rainwater harvesting technique may produce quite dissimilar results in different areas. Particular attention needs to be given to the relationship between soil 9 management and water availability to assure the best possible results from water harvesting operations.

Soil nutrient availability is essential in enhancing the effects of water harvesting and helping to ensure increased yields. The relationship between soil nutrient levels and water harvesting is particularly important in areas of sub-Saharan Africa where soil nutrient levels are generally very low. Increased nutrient availability will also help to promote root development and canopy cover growth, which will increase water uptake by the crops and help to advance biomass growth.

In addition to soil nutrient requirements, the physical structure of the soil also has an impact on the effectiveness of water harvesting. The degradation of the easily erodable soils in many arid and semi-arid regions leads to specific concerns regarding water harvesting methods. The erosion of the sandy surface of these soils, often due to the removal of vegetation by overgrazing or other means, exposes the clayey subsurface that forms a crusty layer with lower infiltration rates. While these crusted surfaces are often abandoned because of their low potential for agriculture, they may prove very useful for water harvesting by inducing runoff from the more impenetrable catchment area to the cultivated area below.

The impact of raindrops on the eroded surface can also help to induce crusting. In some situations, the runoff may also bring nutrient-rich litter along with it to the cultivated area, thus increasing the availability of nutrients to the crops. In cases when a crusty layer is not formed on the catchment area, the soil surface may be treated with other materials to reduce infiltration rates and promote runoff. Some chemical treatments have been used for this purpose, including asphalitic materials and paraffin wax. While these materials have increased runoff efficiency, they are generally only effective for 2-5 years. Other materials such as plastic sheeting, fiberglass and concrete have also been used but these items are often too expensive for farmers in arid areas to afford.

Although the catchment area benefits from a layer of soil with low infiltration rates, this characteristic can be detrimental to crop production in the cultivated area. Very low infiltration rates that result from crusty surfaces can lead to waterlogging in the cultivated area, rendering the area unfit for crop production. The most appropriate type of soil for the cultivated area would be a deep fertile loamy soil. The presence of organic matter improves soil structure and allows for greater water infiltration and better penetration of plant roots. Deep soils are able to hold water from a water harvesting system and may also be able to provide a more nutrients for plant growth.

Water Conservation

Water conservation methods, often referred to as in situ rainwater harvesting, include activities such as mulching, deep tillage, contour farming and ridging. The purpose behind the se methods is to ensure that the rainwater is held long enough on the cropped area to ensure infiltration. These techniques are best suited to areas where rainfall and water holding capacity are sufficient to meet the crop water requirement but the amount of water infiltration is not adequate to reach the required moisture level. Some methods, such as mulching or the addition of organic matter, may also help to enhance the physical characteristics of the soil. Water conservation is often used in tandem with water harvesting techniques in order to achieve better results.

Deep tillage is a water conservation technique that improves soil moisture capacity by increasing soil porosity. In addition, runoff is reduced through increased roughness at the soil surface, which increases the time available for water to infiltrate the soil. This increased infiltration will increase the availability of water in the root zone to assist in plant growth. It is important to note, however, that these techniques are not suitable in all situations.

Soil texture and structure as well as economic limitations that may exist if high capital inputs are needed. For example, draft animal power is essential to deep tillage due to the amount of power needed. In the Dodoma region of Tanzania, few areas use deep tillage techniques because the draft animal power is not available. Contour farming is a technique in which tilling and weeding are done along the contours to help stop water runoff. Mulching or the addition of other organic material to the soil is a water conservation method that may both increase soil water availability by increasing soil water holding capacity and decreasing evaporation and improve the quality of the soil.

The addition of organic matter is often used along with other conservation methods to help increase water infiltration. When organic material is added with conservation tillage systems instead of using traditional mould-board ploughs, the decrease in erosion is even greater. These benefits are further increased when organic material is added to the soil. Beet found that in Ghana, a decrease in organic matter in soils from 5 percent to 3 percent decreased soil water retention from 57 percent to 37 percent. While animal traction may be useful in increasing soil aeration and infiltration of water, it may also result in soil degradation in the form of increased erosion. The development of new farming techniques may be useful in combating these additional problems.

Microatachments

Microcatchment water harvesting systems consist of a distinct catchment area and cultivated area that are adjacent to each other. Some advantages of microcatchments include the high specific runoff yield compared to larger catchments and their simplicity, inexpensiveness and easy reproducibility. Microcatchments have been used in Asia, Africa, America and Australia.

Bunding and v-shaped catchments

Various forms of bunding and v-shaped microcatchments have been used successfully in some arid and semi-arid regions. Contour bunds are earth, stone or trash embankments placed along the contours of the hillside in order to trap rainwater behind them and allow for greater infiltration. Microcatchments using contour bunds were found to be successful for fruit tree plantations in Syria. At least twice yearly runoff was found to be sufficient for the selected tree species (almonds, pistachios, figs and grapes) used in the study. The same method was found to be uneconomical for mountainous areas as the construction costs were too expensive to offset any additional gains in yield.

Semi-circular bunds are generally placed in a staggered formation and allow water to collect in the hoop for greater infiltration. Excess water is displaced around the edges of the bund when the hoop area is filled with water. Contour bunds are generally used on slopes less than five percent, while semicircular bunds are usually only used if the slope is less than three percent.

V-shaped microcatchments are similar to semicircular bunds except that a vshaped catchment area is used instead of a hoop shaped area. A study conducted in Niger used v-shaped microcatchments in the production of millet and sorghum. The catchments were set up so that runoff would collect in the v-shaped area and then overflow around the sides once the basin was filled.

Meskat-type systems

Other types of microcatchments use a catchment area that diverts runoff water directly onto a cultivated area at the bottom of the slope. The meskat-type system differs from those previously described in that the field is divided into a distinct catchment area that is located directly above the cropped area instead of alternating catchment and cultivated areas. The cultivated basin is surrounded by a ushaped bund in order to hold the runoff. Meskat-type systems are used for most types of cereal crops including maize, sorghum and millet.

The Northwest Frontier Province of Pakistan using water catchment aprons of various lengths and slope gradients to examine increases in soil moisture. In this method, a flat cultivated area is located in between two apron catchment areas. They found that soil moisture is significantly increased when aprons of 4 to 5 meters with 7 to 15 percent gradients are used. Moisture was increased by 59 percent in the first 15 cm of soil, by 63 percent in the second 15 cm, and by 80 percent in the 30 to 45 cm depth range. A study in Balochistan, Pakistan compared wheat yields for three experimental fields: a control area in which the entire area was planted, a water harvesting area with a ratio of 1:1 between the catchment area and cropped area, and a second water harvesting area with a ratio of 2:1 between the catchment area and cultivated area.

External catchments

External catchments or macrocatchment rainwater harvesting entails runoff collection from a large area located a significant distance from the cultivated area. The collected water is sometimes stored in a separate location before being used. Some types of external catchments include hillside sheet or rill runoff utilization, floodwater harvesting within a streambed, hillside conduit systems and ephemeral stream diversion.

Hillside runoff and conduit systems: The hillside sheet or rill runoff system uses runoff from a hilltop or other areas, which then collects in flat areas where it is used for cultivation. The runoff is often used without any additional management. In some instances, bunding is also used in order to hold the runoff in the cultivated area. Maize, sugarcane and vegetables are often grown using hillside sheet or rill runoff systems. One of the predominant characteristics of external catchments is the low labor cost since there is generally no flood control management.

Hillside conduit systems are beneficial in areas where the runoff must travel over a long distance before reaching the cultivated area. When the slope along which the runoff must travel is very long, the velocity of the flow is often rather slow, which makes it quite possible that a significant amount of the water may be absorbed before it reaches the cultivated area. In this situation, it is useful to dig channels on the hillside, which increases the runoff flow velocity and allows more of the runoff to reach the field and be absorbed there.

Floodwater harvesting and stream diversion: Floodwater harvesting within a streambed involves blocking the water flow, causing water to concentrate in the streambed. The streambed area where the water collects is then cultivated. It is important to make sure that the streambed area is flat with runoff producing slopes on the adjacent hillsides and that the flood and growing seasons do not coincide. A terraced wadi system is one type of floodwater harvesting, in which a series of low check-dams are constructed across a wadi or ephemeral stream and the wadi area is cultivated. Some important factors to consider when determining if this method is suitable for an area include the soil quality and depth in the wadi and the ratio of catchment area of the wadi and the size of the area to be cultivated in relation to the rainfall runoff.

Ephemeral stream diversion is another external catchment system that is often used to harvest rainwater. In this technique, the water in an ephemeral stream is diverted and applied to the cropped area using a series of weirs, channels, dams, or bunds. There are two main ephemeral stream diversion systems. The first method uses a weir to divert the stream water into a cultivated area close to the stream that has been divided into several open basins using some type of bunding. Once one basin has filled, the overflow discharges into an adjacent basin. The second method uses a weir and a system of channels to divert the water into a rectangular basin. This method can be used on fields that are further away from the stream source.

Water harvesting was found to improve sorghum yields most during seasons with inadequate or poorly distributed rainfall. Additional gains were found when fertilizer was utilized with water harvesting; in the 1987/88 season, yields were 110 percent higher due to water harvesting and yields increased 20 percent due to the addition of phosphorous and manure.

Benefits of Water Harvesting Techniques

While many water harvesting case studies have shown increases in yield and water use efficiency, it is not yet clear if the widespread use of these technologies is feasible. Construction and maintenance costs of water harvesting systems are very important in determining if a technique will be widely adopted at the individual farm level. Additionally, extension and educational support to farmers is crucial to assure that water harvesting methods are adopted and maintained.

Several factors that influence the cost of catchment construction include labor and maintenance costs, soil characteristics, and the size and shape of the catchment.

Labor and construction constitute the bulk of water harvesting costs in areas where ample land is available to construct the water harvesting structures. In land-scarce areas, the use of land for water harvesting will involve an opportunity cost, which may influence a farmer's decision to adopt water harvesting techniques. The initial high labor costs of building the water harvesting structure often provide disincentives for adoption. The initial labor costs for construction generally occur in the dry season when labor is cheaper but also scarce due to worker migration; maintenance costs, on the other hand often occur in the rainy season when labor costs are higher due to competition with conventional agriculture.

While labor costs may be somewhat high in countries with less manpower, countries such as Pakistan with ample labor supplies are expected to be able to construct the systems for a much lower price. The construction costs of small-scale water harvesting systems are quite small when compared to traditional large water projects - one estimate indicates that they are only 50-65 percent of the large-scale construction costs on a per unit cultivated area basis.

Fertilizer costs are also important for establishing water harvesting systems. Due to the link between soil nutrients and water requirements, availability of soil nutrients is essential to the initial development of crops. Animal manure is probably the best option for most farmers in arid and semi-arid areas due to the high cost and lack of access to chemical fertilizer.

Although many rainwater harvesting techniques work well at the experimental level, ensuring on- farm adoption proves to be an additional challenge in testing the techniques on a broader scale. Collective action issues may prove to be a problem when trying to implement water harvesting techniques. Since many of the water harvesting structures are large and require a substantial amount of land and labor to implement, the creation of these structures is often undertaken at a communal level. Problems develop at this level, however, as farmers are often not willing to supply voluntary labor to build the structures and the maintenance of communal water harvesting systems is often neglected.

Factors that may ensure the acceptance of techniques at the farm level include the involvement of farmers from the planning stages and the use of farmers for maintenance and data collection which can create a sense of ownership over the project. In addition, it is useful to provide information about benefits of the water harvesting technology early on in the adoption process to help promote adoption through the provision of appropriate educational and extension support to ensure that the farmers have the knowledge necessary to implement the chosen technology.

The environmental effects of water harvesting for agriculture should also be considered when determining whether to adopt a certain technology. Some examples of environmental damage that may occur due to water harvesting include salinization, sodification, low water tables or water logging, and soil degradation. An additional consideration is how the use of additional rainwater from rainwater harvesting techniques will affect water users downstream that may rely on the same water supply for their crop production.

Methods of Modern Farming

Conservation tillage measures such as minimum till and no till have been tested in some developing countries to conserve soil water and decrease the rate of soil water evaporation. Precision agriculture, which has been used in the United States, has also been suggested for use in developing countries. Along with research on integrated nutrient management, applied research to adapt conservation tillage technologies for use in unfavorable rainfed systems in developing countries could have a large positive impact on local food security and increased standards of living.

Conservation and no-tillage techniques may be especially helpful in areas where farmers do not have the capital or labor required for other techniques. The usefulness of these methods will also depend, however on the soil texture and structure. Trials of some conservation tillage techniques have been undertaken in areas of Sub-Saharan Africa with mixed results. There are three seedbed preparation techniques (tillage in dry soil conditions, scarification and different sizes of sweeps, and direct seeding with no tillage) in dry soil for better soil-water management.

Tillage in dry soil conditions is not sustainable over time and should not be recommended to farmers, as it can promote erosion from wind and first rain events. Scarification of the soil surface allows for protection against erosion and runoff due to the addition of crop residues to the soil. This method is used in the semi-arid region of the Sahel but specific results on yield increases have yet to be proven. Direct seeding in dry soil has been used in groundnut production in Senegal. Advantages to this method include: reduced production costs, diminished soil erosion, decreased runoff, less soil compaction, and better timeliness in seeding.

No-tillage technology (often used with mulching) has been found to improve soil moisture conservation and thus reduce crop failure in dry years, particularly in arid or semi-arid areas such as Sub-Saharan Africa. Additional soil improvements such as enhanced soil structure and increased organic matter content have also resulted. In many instances the technology is very location specific and requires high levels of investments for tailoring it to specific conditions and to disseminate it to particular groups of farmers.

Precision agriculture has had some success in the United States and other developed countries. While traditional agricultural techniques have tended to apply the same management to an entire field, precision agriculture methods focus on information technology using site-specific soil, crop and other environmental data to determine specific inputs required for certain sections of a field. Many of these methods involve the use of technologies such as geographic information systems (GIS), satellites, and remote sensing.

Precision agriculture can directly increase crop yields, and also improve water availability through greater relative infiltration of rainfall. In developing countries, the smaller farm sizes could allow for management on a field basis. Precision agriculture may hold significant promise in the future for agriculture in developing countries, as nutrient levels can vary greatly from field to field.

For example, a recent study in the Philippines showed that variations in rice yields from 2400 to 6000 kg/hectare in 42 different fields were attributed to differences in soil nitrogen. More accurate analysis of soil nutrient levels could assist farmers in determining fertilizer levels specific to different areas in the field.

Supplemental Irrigation

Supplemental irrigation is another method used in low rainfall areas to assure that crops receive enough water in order to survive. While water harvesting is generally used in areas that receive between 100 - 300 mm of rainfall annually, supplemental irrigation is used in areas with a slightly greater annual rainfall of approximately 300 - 600 mm. Supplemental irrigation has been described as a technique used on crops that can be grown using rainfall alone, which applies a limited amount of water during times of low rainfall to ensure that enough water is received to support crop growth and stabilize yields. The goal of supplemental irrigation is to provide enough water during critical growth stages to produce optimal yield per unit of water, not to provide stress-free conditions throughout the growing season with the aim of producing maximum yield.

This differs from conventional irrigation in that the amount of water applied in supplemental irrigation would not by itself be sufficient to ensure crop growth. Conversely, conventional irrigation supplies the entire water needs to the crop because rainwater may not provide sufficient water for plant growth for all or part of the season. Conventional irrigation is used in regions where water is plentiful, while supplemental irrigation is often used in places where water is often scarce.

Timing of water application is one of the most important factors to be determined when using supplemental irrigation. Supplemental water applications are especially important when water is scarce during critical growth periods. Oweis, Hachum and Kijne present a rule of thumb that for crops such as alfalfa, maize and spring grains, supplemental irrigation is needed when the soil water content drops to a rate of 50 percent of available water (which is equal to the field capacity minus the permanent wilting point) in the root zone. Other crops such as potatoes and vegetables will produce better when soil water is kept within the top 35 percent of available water.

The water used for supplemental irrigation can be obtained from a variety of sources. Groundwater, surface water, industrial wastewater, and water obtained through water harvesting methods are all used for supplemental irrigation. The water harvesting methods discussed earlier are often used in conjunction with supplemental irrigation since SI is often undertaken in low-rainfall areas. Important factors to consider when designing a water harvesting system for supplemental irrigation include the storage capacity, type of storage and storage location.

Specific methods of irrigation used depend upon the resources available to the farmers in an area as well as any economic or labor costs that may be involved with setting up the SI system. Potential benefits that can be achieved through the use of supplemental irrigation include increased yields, stabilization of yields across years, and creating conditions that allow for the use of higher technology inputs such as high- yielding varieties, herbicides and fertilizers.

Rainfed Cereals

A common perception is that rainfed areas did not benefit much from the Green Revolution, but breeding improvements have enabled modern varieties to spread to many rainfed areas. Over the past 10-15 years most of the area expansion through the use of modern varieties (MVs) has occurred in rainfed areas, beginning first with wetter areas and proceeding gradually to more marginal areas. In the 1980s, modern varieties of the major cerealsspread to an additional 20 million hectares in India, a figure comparable to adoption rates at the height of the Green Revolution. Three quarters of the more recent adoption took place on rainfed land, and adoption rates for improved varieties of maize and wheat in rainfed environments are approaching those in irrigated areas.

The adoption rate of modern varieties of the major cereals varies by cereal type and region. In rainfed areas of developing countries, the spread of wheat MVs (Type I change) lead to a 15-20 percent yield gain over traditional varieties (TVs), while annual yield gain attributed to adopting newer generations of MVs (Type II change) averaged 0.5-1.0 percent in high rainfall areas with almost

no gains in very dry areas in the past thirty years. Compared to extensive diffusion in the United States where hybrid maize covered 96 percent of the maize area by 1960, a little less than half of the maize area in developing countries is sown to MVs (hybrids and open-
30 pollinated varieties), ranging from 36 percent in sub-Saharan Africa to 66 percent in East, South, and Southeast Asia.

Adoption of MVs in upland rice ecosystems has been discouraged by the poor growing conditions and poverty levels, and there has been zero use of MVs in flood-prone environments. Improved varieties of secondary cereals and other grains such as pulses that are grown widely in marginal rainfed environments may also increase production. In Africa, sorghum and millet are grown in a harsh semi-arid tropical climate where inadequate rainfall and lack of irrigation make production of other cereal crops difficult to sustain.

The adoption of new varieties of sorghum and millet is likely to have a very small impact on yields unless there is also a fairly rapid increase of input use, especially inorganic fertilizers and quality seeds. However, with appropriate input use, improved millet varieties are estimated to increase yields by 22 percent or about 0.2-0.5 ton per hectare even under the dry conditions of Sahelian countries like Niger.

Future Improvements in Rainfed Crop Yields

Significant long-term increases in potential crop yield are possible in favorable rainfed areas, but until recently, potential cereal yield increases appeared limited in the less favorable rainfed areas with poor soils and harsh environmental conditions. However, recent evidence shows dramatic increases in yield potential in even drought-prone and high temperature rainfed environments. A change in breeding strategy to directly target rainfed areas, rather than relying on "spill- in" from breeding for irrigated areas was a key to this faster growth.

Past breeding strategies for both irrigated and rainfed areas emphasized yield maximization. For example, the maize seed presently distributed comes from varieties selected without sufficient attention to agronomic and socioeconomic factors that limit production: periods of drought and/or frost during the crop's vegetative stage, soil exhaustion, grain type and color, and forage qualities. More recent crop genetic improvements through both conventional and non-conventional breeding have placed more of an emphasis on altering specific characteristics of the crop.

Conventional breeding uses whole plants to select desired characteristics. These techniques have generally focused on maximizing yields through increased plant productivity and resistance to stress. Non-conventional breeding uses cellular and molecular biology techniques such as marker-assisted selection and cell and tissue culture techniques, which allow plants to be screened more quickly in the laboratory rather than the field. These techniques can also more rapidly and efficiently select for particular characteristics that may increase tolerance to diseases, pests, and adverse weather conditions, thus increasing yields. Non-conventional breeding also inc ludes the transfer of genetic material from one species into another, creating transgenic or genetically modified organisms (GMOs) with beneficial characteristics that cannot be achieved through conventional breeding.

Both conventional and non-conventional breeding techniques are used with the goal of increasing yields. Three major strategies that are used to increase yields include an increased harvest index (HI), an increase in the general plant biomass, and increased stress tolerance (particularly drought resistance). The first two methods increase yields by altering the plant architecture, while the third focuses on increasing the ability of plants to survive stressful environments.

Harvest Index (HI)

One method to improve yield is to increase the harvest index (HI), but this technique may have only limited potential for generating further yield growth. In recently released varieties, the HI of rice is about 0.50-0.55, and the scope for continued increases is limited by the need to maintain sufficient leaf area and stem biomass for interception of solar radiation, physical support, and storage of assimilates and nitrogen used in grain filling. International Rice Research Institute (IRRI) estimates that it will not be available to farmers until 2005 at the earliest and its adaptability to rainfed environments remains unclear. Recent wheat cultiva rs appear to have a relatively low HI of 0.41-0.47 when grown with irrigation in California and Mexico but a further increase in HI might be feasible.

Total Plant Biomass

Another strategy to increase yield is to increase the total dry matter (biomass per unit of land and inputs such as water and nitrogen) by increasing plant productivity. The development of hybrid varieties and increased photosynthesis are both techniques that focus on increased biomass.

Hybridization

Hybrid varieties increase plant productivity and provide greater yields than conventional varieties, making use of heterosis (or hybrid vigor), the phenomenon in which the progeny of two distinctly different parents grow faster, yield more, and resist stress better than either parent. Hybrid varieties of all of the major cereals have been developed throughout the past several decades. An overview of some of the major research results follows.

Hybrid maize was the first hybrid to be developed among the major cereals since it is a cross-pollinated species. Attempts to introduce U.S. hybrids in subtropical and tropical environments during the 1950s failed due to the lack of adaptation to climate, disease, and insect pests. However, later improvements in hybrid maize have allowed diffusion of hybrid maize in subtropical and tropical regions.

It is 35 estimated that nearly half of the maize area in developing countries is sown to hybrid maize, while improved open-pollinated varieties (OPVs) occupy a lower proportion of around 20 percent. The substantial adoption of hybrid maize in developing countries shows that the cost of purchasing hybrid seeds each planting season is not prohibitive for many farmers compared to the income benefits of these varieties.

The adoption of improved varieties has had a significant impact on maize production in Africa. For hybrids, most estimates suggest that on-farm yield gains have averaged at least 40 percent in favored areas, with dry areas or drought years providing at least a 30 percent yield gain.

According to Morris, Clancy, and López-Pereia, the yield gain due to improved OPVs is probably less, with a yield gain of about 15-25 percent compared to local materials in tropical areas. Applying these data to the estimated area planted to MVs in Africa, the overall yield gain due to adoption of MVs alone is 12-14 percent. Maize area expanded as well, especially in the drier savanna areas, since early maturing varieties and hybrids became available.

Although a male sterile/fertility restorer systemin wheat was discovered in the early 1960s, hybrid wheat has not been regarded as a viable option until recently. Wheat is a self-pollinating plant like rice, requiring more complicated and costly processes to obtain male sterility than cross-pollinating plants such as maize. Cytoplasmic 36 male sterility, nuclear male sterility and chemical hybridizing agents (CHAs), which are chemical agents used to produce male sterility, are three systems leading to male sterility in wheat. Due to the slow process of cytoplasmic male sterility that lacks versatility in methods of fertility restoration and the lack of cost-effective methods of maintaining nuclear male sterility, several companies abandoned their hybrid wheat research and development programs. Those who continued rescheduled their strategy to include CHAs and biotechnology.

However, hybrid wheat varieties based on cytoplasmic male sterility/restorer systems have been developed, produced and marketed. In South Africa, wheat hybrids tested in 25 locations in the Free State in South Africa out- yielded the non- hybrids tested by 14.8 percent and 11.5 percent in 1994 and 1995, respectively. These hybrids also have an advantage over conventional cultivars under conditions of water stress. Given these yield and water stress tolerance advantages, hybrid wheat technology may be of particular importance for wheat production under marginal conditions.

Although most hybrid rice has been bred for irrigated environments, the possibility of spreading the technology to rainfed areas may hold potential to increase rainfed yields. Since the release of the first commercial Fhybrid rice in 1976 in China, Fhybrid rice was planted on about 18 million hectares of the total 33 million hectares planted to rice by 1992. Those hybrid varieties yielded about 20 percent higher than inbred varieties. Hybrid rice technology has been limited to temperate and subtropical regions; however, tropical hybrids have recently become available in other Asian countries.

Vietnam, India and Bangladesh plant hybrid rice on 200,000, 150,000, and 30,000 hectares respectively, reporting yield increases of up to 20 percent above the best yields of semidwarf rice varieties. Hybrid rice between indica subspecies increased yield potential by about 9 percent under tropical conditions. This gain is attributed to the greater biomass production rather than harvest index. The diffusion of hybrid rice has slowed in China because of the lower eating quality of grains. There has so far been limited success in developing high- yielding cultivars that also have better eating quality.

Further enhancement in yield potential may be possible from the use of intersubspecific heterosis because the magnitude of heterosis is higher if the genetic difference in the parents is greater. Indica and japonica rice germplasm have, as opposed to the narrowed genetic diversity among the improved indica rice due to massive international exchange of germplasm, remained distinct, as there has been very little gene flow between these two varietal groups. As expected, the hybrids between indica and japonica parents showed higher heterosis for yield. It has been shown that the level of heterosis in the indica/ tropical japonica hybrids is higher than that of indica/ indica hybrids.

Increased Photosynthesis

Increased photosynthesis or radiation use efficiency aims to improve carbon accumulation productivity, contributing to greater biomass production. One example of an attempt to increase photosynthesis is research to create Crice plants. Tropical species such as maize, sorghum, and sugarcane are Cplants, which have evolved a more efficient photosynthesis mechanism than Cplants such as rice, barley, and wheat.

Cplants may lose up to 50 percent of their recently fixed carbon through photorespiration, while the carbon loss of Cplants is greatly reduced or completely inhibited; therefore, C4 plants are more advantageous in their photosynthetic productivity. It should be noted, however, that Cspecies do not always have higher water use efficiency. Under non- ideal conditions such as drought, there are more tolerant Cspecies such as cowpea and cotton than the comparatively sensitive Cmaize and sorghum cultivars.

The potential benefits from modifying rice, a Cplant, by incorporating traits currently found in Cplants such as maize, sorghum and sugar cane, are large. While the rice plant already contains all of the genes responsible for Cphotosynthesis, they are not switched on and regulated as they are in maize. By transferring genes with an improved mechanism for the process of photosynthesis from maize to rice, researchers in the United States and Japan (Tsukuba) have produced initial results that suggest that rice yields could be increased by up to 20 percent.

Functional genomics increases understanding of drought tolerance control and defines strategies for crop improvement. In several cereal species, genetic maps have helped to identify chromosomal regions controlling some traits related to drought stress response. Cereal crops such as maize, sorghum, rice, wheat and barley have been studied to identify regions controlling 42 characteristics such as phenology, root characteristics, plant architecture and growth, photosynthesis, chlorophyll amount or "stay green" character, and water-use efficiency.

The limiting factors in rainfed wheat production are drought, temperature extremes, low nitrogen, low phosphorus, soil acidity, high aluminum saturation, and